Consciousness-Based Evolution

John S. Torday

Professor of Pediatrics,
Obstetrics and Gynecology
Evolutionary Medicine
University of California-Los Angeles
Los Angeles, California

Director
The Henry L. Guenther Laboratory for Cell-Molecular Research
The Lundquist Institute
Harbor-UCLA Medical Center
Torrance, CA

Fellow
The European Academy of Science and Arts

CRC Press
Taylor & Francis Group
Boca Raton London New York

CRC Press is an imprint of the
Taylor & Francis Group, an **informa** business

A SCIENCE PUBLISHERS BOOK

First edition published 2024
by CRC Press
2385 NW Executive Center Drive, Suite 320, Boca Raton FL 33431

and by CRC Press
4 Park Square, Milton Park, Abingdon, Oxon, OX14 4RN

CRC Press is an imprint of Taylor & Francis Group, LLC

Library of Congress Cataloging-in-Publication Data (applied for)

ISBN: 978-1-032-19701-2 (hbk)
ISBN: 978-1-032-19702-9 (pbk)
ISBN: 978-1-003-26040-0 (ebk)

DOI: 10.1201/9781003260400

Typeset in Palatino Linotype
by Radiant Productions

Dedication

I would like to dedicate this book to the people that I love, namely my wife Barbara J. Torday, my daughter Nicole A. Torday, my son Daniel P. Torday, my daughter-in-law Erin Sullivan Torday, and my grandchildren Abigail Eve Torday and Delia Rose Torday.

I would also like to dedicate this book to my mentors Dr. Claude Giroud (McGill University), Dr. Mary Ellen Avery (Harvard University), Dr. N.L. First (University of Wisconsin-Madison) and Dr. Jack Gorski (University of Wisconsin-Madison).

Preface

The consensus is that the most important thing we need to understand is consciousness, which has formally eluded us since the time of the Ancient Greek philosophers. Our search for consciousness has focused on the brain as the source of mind, but that has not borne out, resulting in the 'explanatory gap'- the failure to connect that wet 3 pounds of brain to the 'voices in our heads'. The premise of this book is that the only way we can and will ever understand consciousness is through Science and the Scientific Method because it is the only means for 'knowing what we don't know'.

In that vein, I have spent my career trying to understand the causal nature for our physiology. In doing so, I began systematically reverse-engineering cellular physiologic development about 25 years ago based on cell-cell communication as the only mechanism we know of (to date) for generating form and function.

Beginning with the development of the lung, which is my specific field of interest, I exploited what was known about the paracrine mechanisms for the formation of various physiologic tissues and organs, beginning with the fertilized egg, which itself starts the cell-cell signaling cascade between the animal and vegetal poles.

The big breakthrough in my own study of physiologic evolution was the serendipitous observation that deleting Parathyroid Hormone-related Protein (PTHrP) during embryonic mouse development resulted in failure to form alveoli, resulting in death shortly after birth. Subsequent study of these PTHrP-deficient offspring revealed more subtle failure of skin, kidney, bone and brain development, revealing the pleiotropic nature of this gene. Its primary feature is that it mediates the effect of 'stretch' on physiologic homeostasis in all of these organs and tissues, raising the question as to what the 'common denominator' for all of these adaptations is? There were three hormone receptor duplications that occurred during the water-to-land transition 500 million years ago- the PTHrP Receptor, Glucocorticoid Receptor and ßAdrenergic Receptor. Given the primacy of adapting to the increased force of gravity (from buoyancy to land), the amplification of PTHrP signaling was necessary in order to remodel the skeleton, specifically the evolution from fish fins to tetrapod legs. Those members of the fish species that were most able to up-regulate their PTHrP signaling did so in other organs as well, facilitating adaptation to land- namely, the evolution from the fish glomus to the mammalian glomerulus, and the barrier function of the skin.

Since astronauts develop osteoporosis due to decreased tension on bone in response to microgravity, I tested the hypothesis that PTHrP mRNA 'expression' would decline in zero gravity. Lung and bone cells were exposed to 'free fall' as a proxy for zero gravity, causing a significant decrease in PTHrP mRNA over the course of 8 hours, reaching a new baseline. Since PTHrP is critically important in the evolution of the lung and bone, the loss of its expression suggests that the cells had lost their evolutionary 'history'. And because vertebrate evolution has been mediated by symbiogenesis- the assimilation of existential threats in the environment- it raised the question as to where symbiogenesis emerged from. Reasoning that symbiogenesis is in service to homeostasis, and that Quantum Mechanics underlies all of existence, it was hypothesized that Quantum Entanglement, which maintains balance among electrons and photons, was the foundation for symbiogenesis, which is essentially the mechanism for maintaining homeostatic balance. And since our physiology mediates our consciousness, gravity is the force that generated the latter.

It appears as though our primary 'purpose' is to detect novel changes in our environment and evolve in response, based on epigenetic inheritance. The following provides the rationale for that understanding.

Acknowledgements

I have been supported by research grants for my entire career, freeing me to do the experiments that are largely the basis for this book. Among them have been The National Heart, Lung and Blood Institute; The March of Dimes; The Thrasher Foundation; The American Heart Association; The Hood Foundation; The Parker B. Francis Foundation; The Peabody Foundation; The Meadows Foundation; The Edward Beatty Foundation; The King Trust.

Thank you all.

Contents

Dedication   iii

Preface   v

Acknowledgements   vii

1. Evolutionary Basis for Consciousness, Subconsciousness   1
   and Unconsciousness

2. Physiology as Consciousness   5

3. It's Epigenetics 'All the Way Down'   16

4. Ontogeny, Phylogeny and Cellular Energy Flows for Evolution   20

5. Cellular Evolution of Language as Quantum Energy   33

6. Fibonacci Numbers, Self-Reference, Self-Organization   47
   and Autopoiesis

7. Jung's Synchronicity, Unicellular Consciousness,   52
   Quantum Mechanics and Epigenetic Inheritance

8. Life is a Simulacrum of Cosmologic Physics, Mathematics,   56
   Art, Music

9. Morphological Forms Emerging from the Evolutionary   65
   Process are Topologies

10. On the Quantum Origin and Nature of Consciousness   74

11. The Holism of Cosmology and Consciousness   86

12. Evolution, Gravity, and the Topology of Consciousness   94

13. Cellular Evolution and the Flow of Energy   103

14. Fractal Properties of Physiology — How and Why   112

15. Embodied Quantum Entanglement   127

16. If Music Be the Food of Love, Play On....   136

17. Implicate and Explicate Orders as Unconscious and Conscious   141

18. The Cell as a Mobius Strip   161

19. Two Paranormal Psychologists Walk into a Bar   168

Index   **173**

Chapter 1
Evolutionary Basis for Consciousness, Subconsciousness and Unconsciousness

Introduction

In the book "Hormones and Reality" (Torday, 2022) consciousness was framed in the context of how the endocrine system might explain such phenomena. This was of particular interest because of the discovery that the endocrine system is under epigenetic control, raising the question as to 'how and why' that might aid the evolutionary process. In an earlier work, the evolution of human consciousness was addressed from the perspective of bipedalism emerging from warm-bloodedness, freeing the forelimbs for toolmaking and language, placing great positive selection pressure on the evolution of the central nervous system. The anatomic evidence for this functional interrelationship emanates from the Area of Broca, an area of the brain in the frontal cortex that is exclusive to Great Apes. It is where the capacities for language and toolmaking emanate from. The combination of toolmaking and language is a powerful catalyst for creativity—the synergy between imagination and toolmaking marks the ascent of human intelligence, particularly in combination with bipedal locomotion to facilitate these traits. That is particularly true when considered in the context of the progressive enlargement of the human head over the course of hominin evolution. Eventually, the infant could no longer fit through the birth canal, so human infants are born prematurely with respect to the development of the brain, having only about 25% of adult brain volume at birth. It takes decades for humans to mature to full brain capacity, rendering them highly dependent on social systems like the family, extended family, town, city, and nation. In the meantime, they exhibit immature behaviors like risk-taking, narcissism and the like, which have spawned human social systems to support such otherwise adolescent activities. As a result, we have both great creations of human activity like literature, art, architecture and education. But we also have wars and poverty and economic collapse due to human's lack of constraint, unlike all other species. If only we could maximize our positive traits and minimize our negative traits, we would be better off.

Inherent in the cell-cell communication basis for evolution (Torday and Rehan, 2012) is the 'memory' of the organism's history as the reference point(s) for physiologic change. When confronted with Romer's "Greenhouse Effect", drying up bodies of land and depleting oxygen from the water, boney fish self-selected to transition to land, facilitated by three gene duplications that allowed for the morphing of the swim bladder into the lung (Torday and Nielsen, 2017). This was due to the deep memory of The First Principles of Physiology (Torday and Rehan, 2009), and cell-cell signaling as the basis for development, physiology and injury-repair. In the alveoli of the lung, the glomeruli of the kidney and in bone this translates into stretch-regulated cell-cell signaling as the manifestation of the force of gravity, rendering these structures/functions even more sensitive to changes in the environment as a result.

It has been proposed that life originated from lipid molecules produced by Pulsars in deep space, immersed in the primordial ocean that initially covered the Earth (Zahnle et al., 2010). Such lipid molecules coalesced to form micelles, or prototypical cells. That process was contingent on gravity (Torday, 2003) causing the lipid molecules to orient vertically to the surface of the water, their negatively-charged hydrophilic ends pointing downward into the water. Once a critical mass of negatively charged lipid molecules packed together at the air-water interface, their charge neutralized the Van der Waals force that causes surface tension, the lipid molecules 'quantum leaping' from individual molecules to the micelle, a lipid membrane-bound sphere that is semipermeable. Subsequently, the negative entropy within the micelle (Schrodinger, 1944), fueled by the energy produced by gravitational force on the curved surface of the micelle facilitated Symbiogenesis (Sagan, 1967) as the step-wise mechanism for evolution in response to an ever-changing environment. In other words, the ontology of life is characterized by both the 'holism' of the micelle and the 'part-ism' of Symbiogenesis. The holism is manifested by the cell membrane acting as the topology that separates the inside of the cell from the external environment. It behaves as a semipermeable membrane, making binary decisions about whether any given particle should be inside or outside the cell. Subsequently, such particles were assimilated to form our physiology, controlled by cell-cell communications mediated by soluble growth factors and their cognate receptors.

In keeping with the theme of this book, the fractal basis for Symbiogenesis is in service to homeostasis. And since the consensus is that Quantum Mechanics is the foundation for all existence, the question is which aspect of quantum mechanics sustains and perpetuates homeostasis? Quantum Entanglement (QE) is the property of Quantum Mechanics that qualifies as the homologue for Symbiogenesis as the mechanism for homeostasis. In turn, since QE is characterized as being both local and non-local, it offers a way of understanding consciousness as fractals of its local and non-local characteristics. And given that consciousness has arisen from physiology (Torday, 2020), the fractal nature of consciousness also forms the fractal basis for physiology.

The converse is also true, that when one meditates there is a diminishing of our sense of complexity (Aftanas and Golocheikine, 2002) as we regress consciousness from the brain to the gut to the skin as the reverse of the sequence of brain evolution. Empirically, Mashour (2021) has shown that when patients come out from under general anesthesia they recapitulate the phylogenetic changes in the brain, which is the opposite of the effect of meditation. Similarly, Near Death Experiences are described as being reduced to a point of light (Nelson, 2015). And the Runner's High due to an outpouring of endorphins from the posterior pituitary produces a sense of holism or wellbeing. On another level, this is not unlike looking at a painting, reading a poem, or solving a mathematical problem as purely cerebral processes, feeling a sense of the 'whole' or the 'hole'.

There is evidence for physiologic 'memory' such as the experiments done with planaria, running them through a maze, grinding them up and feeding them to control planaria that are then able to run the maze from memory (McConnel, 1962). Or Phantom Limb, which is a means of remembering the up-stream signaling to avert disuse atrophy. Or Qualia, subjective conscious experiences. Or what David Chalmers referred to as the 'hard problem', why we see red when we whack our thumb with a hammer (Chalmers, 1995). All of the above may be due to memories embedded in our physiology which we elicit under various conditions unconsciously that then become conscious. The underlying mechanism is hypothetically due to the cellular interconnections that are conventionally seen as our physiology (Torday, 2020), but which also can reprise our evolutionary history. This is not unlike the experience we have during meditation as an active process for doing the latter, unconsciously deconvoluting memories in our brain, gut and skin, the latter two being atavistic origins of mind.

This dual system forms the basis for our understanding of mind, which we describe dually as consciousness and subconsciousness, referencing the aggregate of our physiology, but without a mechanistic understanding of their evolution. There is not a great deal of experimental evidence for such a dual-leveled view of consciousness. The Libet Experiment (1983) is consistent with it, observing a significant time delay of some 500 milliseconds between the brain's reaction to an electrical stimulus and the physical manifestation. The 'two slit experiment' also reveals a dualism of consciousness, the patterns of a beam of light passing through a two-slit template being consistent with both particle and wave—when you look at the template you see a pattern consistent with photons passing through it, whereas if you look away the beam of light behaves like a wave. This phenomenon may be a manifestation of the two-tiered consciousness that was described above.

References cited

Aftanas, L.I. and Golocheikine, S.A. 2002. Non-linear dynamic complexity of the human EEG during meditation. Neurosci. Lett. 330: 143–146.

Chalmers, D. 1995. Facing up to the problem of consciousness. J. Consciousness Studies 2: 200–219.

Libet, B., Gleason, C.A., Wright, E.W. and Pearl, D.K. 1983. Time of conscious intention to act in relation to onset of cerebral activity (readiness-potential)—the unconscious initiation of a freely voluntary act. Brain 106: 623–642.

Mashour, G.A., Palanca, B.J., Basner, M., Li, D., Wang, W., Blain-Moraes, S., Lin, N., Maier, K., Muench, M., Tarnal, V., Vanini, G., Ochroch, E.A., Hogg, R., Schwartz, M., Maybrier, H., Hardie, R., Janke, E., Golmirzaie, G., Picton, P., McKinstry-Wu, A.R., Avidan, M.S. and Kelz, M.B. 2021. Recovery of consciousness and cognition after general anesthesia in humans. Elife 10: e59525.

McConnell, J.V. 1962. Memory transfer through cannibalism in planarium. J. Neuropsychiat. 3(suppl 1): 542–548.

Nelson, K. 2015. Near-death experiences—Neuroscience perspectives on near-death experiences. Mo. Med. 112: 92–98.

Sagan, L. 1967. On the origin of mitosing cells. J. Theor. Biol. 14: 255–274.

Schrodinger, E. 1944. What is Life? Cambridge University Press, Cambridge.

Torday, J.S. 2003. Parathyroid hormone-related protein is a gravisensor in lung and bone cell biology. Adv. Space Res. 32: 1569–1576.

Torday, J.S. and Rehan, V.K. 2009. Lung evolution as a cipher for physiology. Physiol. Genomics 38: 1–6.

Torday, J.S. and Rehan, V.K. 2012. Evolutionary Biology, Cell-Cell Communication and Complex Disease. Wiley, Hoboken.

Torday, J.S. and Nielsen, H.C. 2017. The molecular apgar score: a key to unlocking evolutionary principles. Front. Pediatr. 5: 45.

Torday, J.S. 2020. Consciousness, Redux. Med. Hypotheses 140: 109674.

Torday, J.S. 2022. Hormones and Reality. Springer Nature, Switzerland.

Zahnle, K., Schaefer, L. and Fegley, B. 2010. Earth's earliest atmospheres. Cold Spring Harb. Perspect. Biol. 2: a004895.

Chapter 2
Physiology as Consciousness

Introduction

Much has been written and hypothesized regarding what consciousness is, yet we still do not understand what the term actually means. Chalmers (1995) set the bar when he identified the 'hard' problem, asking why we see red when we whack our thumb?, raising the question as to what 'qualia' are? The problem centers on what is referred to as the 'explanatory gap', the difficulty that physicalist theories have in explaining how material properties give rise to the way things feel when they are experienced (Levine, 1983). And yet it is recognized that the evolutionary process of Symbiogenesis has resolved that problem by assimilating physical factors in the environment to maintain homeostatic balance within the cell, in sync with Quantum Entanglement, characterized as local and non-local. In turn, such factors have been made useful as our cellular physiology, founded on the fractals formed by Quantum Entanglement, inferring that it is our capacity to emulate our environment that constitutes our being, so why not recognize it as our consciousness?

It has previously been conjectured that the consequences of phenotypic agency are that "Those impacts are brought back to the eukaryotic unicell and then, upon reproductive elaboration, enable the reiterative extension of phenotype into the environment to experience a subsequent series of environmental impacts towards its next set of adjustments". It is this consistent reciprocation of our physiology with our environment that shapes phenotype. Only once we acclimate ourselves does a newly-acquired trait become constitutive. That process may be natural or artificial. For example, up to a point, adaptation to cold weather can be tolerated, but once the climate becomes intolerable wearing clothes and constructing homes with heat are necessary. Similarly, we have been able to adapt to breathing under water using Self-contained Underwater

Breathing Apparatus (SCUBA), or to outer space using pressurized suits equipped with an oxygen supply—It is only once this is fully considered that a new concept becomes a novel and testable route for further progress in evolutionary theory and biomedicine.

As such, it is this awareness of our surroundings through our senses that actually constitutes consciousness as a robust way in which we are simultaneously aware of both the real-time synchronic and deep historic diachronic aspects of our life. And as we adapt through the loss and subsequent re-establishment of homeostasis by modifying the cell-cell signaling that forms the basis for both synchronic and diachronic consciousness with reference to the First Principles of Physiology at the origin of life, self-referentially self-organizing in order to conform to the everchanging environment is what we conventionally refer to as evolution. It is that coordination of our initial conditions—negative entropy, chemiosmosis and homeostasis—with the on-going surveillance of the environment, combined with the modification of physiology through the acquisition of epigenetic marks that constitutes evolution.

Upon further reflection, that process is constituted by a dual-levelled mechanism that accommodates both the holism of the micelle and its on-going modification through symbiogenesis to remain in sync with the ever-changing environment of an expanding Universe. Indeed, the notion of both consciousness and sub- or unconsciousness are not novel, but the understanding of their cellular origin and coordinated function are.

Freud, Jung, James, ...

In the aggregate [see Figure], this way of understanding the processes of development and phylogeny as cell-cell signaling [Figure, schematic on the left side], differing in time scales, but with the same underlying mechanism, provides an integrative way of understanding the purpose of life as a 'means' for life instead of an 'end'. It is in that sense that life actually constitutes the flow of energy [see Figure] in the former rather than in material being. The validity of this otherwise unconventional view can be defended using Occam's Razor, the simplest answer being the correct one. Darwin's 'entangled bank' describes how "It is interesting to contemplate a tangled bank, clothed with many plants of many kinds, with birds singing on the bushes, with various insects flitting about, and with worms crawling through the damp earth, and to reflect that these elaborately constructed forms, so different from each other, and dependent upon each other in so complex a manner, have all been produced by laws acting around us." In contrast to that, following the flow of energy is much simpler, like following the flow of water, largely eliminating the superficial differences between organisms.

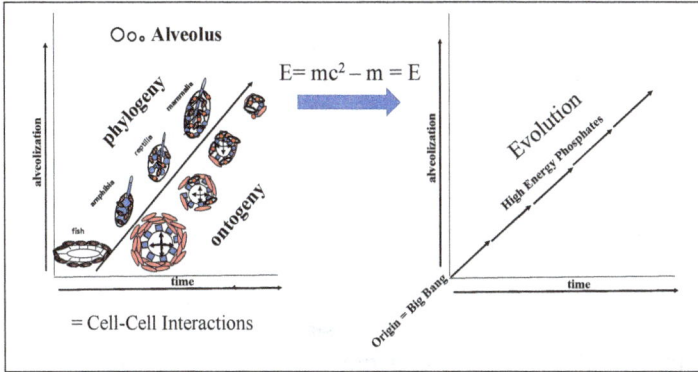

Figure. Evolution, from matter to energy. The phenotypic evolution of lung ontogeny and phylogeny based on cell-cell communication is depicted on the left. Evolution as the flow of high energy phosphate 'second messengers' is depicted on the right. Occam's Razor is more consistent with the flow of energy.

Darwinian evolution is focused on material competition between organisms, whereas cellular epigenetic evolution is predicated on cell-cell communication of from one stage of life to another-developmentally, phylogenetically, as injury-repair ultimately governed by the First Principles of Physiology (Torday and Rehan, 2009). This mode of evolution is founded on ontogeny as the only known mechanism for the formation of biologic structure and function. In turn, Ontogeny is the mechanistic foundation for phylogeny as speciation (Torday and Rehan, 2012, 2017). Merging ontogeny and phylogeny as reducible to cell-cell communication mediated by soluble growth factors and their receptors yields evolution acts as one unified process (Torday and Rehan, 2012, 2017). In that vein, the material aspects of the organism can be seen as 'means', not 'ends', begging the question as to what the 'ends' of evolution are? By eliminating the material aspects of lifeforms, only the flow of energy remains, within and between generations.

That said, what are the ontologic and epistemologic significance of evolution when seen as energy flows rather than material change?

The Big Bang as the vectoral flow of all things

The earth is about 4 billion years old, having formed from the debris of the Big Bang, which occurred about 13.8 billion years ago (Hawking, 1988). The discovery of such a cataclysm is based on the Redshift, the infrared radiation detected by Penzias and Wilson (1965), which was a huge breakthrough in our understanding of how and why the Cosmos formed from a source, the Singularity, prior to which all that existed were descriptions of various celestial phenomena without a central theory of

Astronomy. The astrophysicist Lee Smolin has exploited Darwinian evolution in order to explain the formation of Black Holes and stars (Smolin, 1999), for example. More importantly, the existence of an 'origin' for the Cosmos infers that there is a vectoral 'beginning', not unlike the origin of life as a unicell (Torday and Rehan, 2012; Deamer, 2017), the immersion of lipids in water spontaneously generating micelles, providing a beginning for life.

Given that the Cosmos began as the explosive disruption of the Singularity (Hawking, 1988), releasing energy, homeostasis is the 'recoil' of that event, every action having an equal and opposite reaction (Newton's Third Law of Motion); matter formed as a consequence of that homeostatic counterforce, but energy is the primary state of being (Whitehead, 1978). Therefore, it would behoove us to reconsider evolutionary change as energy instead of matter.

Complexity: a material Artifact of evolution

Cell-cell communication is the basis for forming complicated phenotypes mediated by high energy phosphates as 'second messengers'. Such messengers affect the transcription of DNA within the nucleus, leading to translation to RNA and protein, which is the 'central dogma' of biology (Crick, 1970) the proteins affecting either growth or differentiation of the affected cell. The targeted cell then produces soluble growth factors that affect other neighboring cells, causing either growth or differentiation, etc. We tend to focus on the material aspect of this cascade of developmental forms, and ignore the intermediate transfer of energy from cell to cell, considering it to be a means, and not an end. However, as indicated above, with the merging of ontogeny and phylogeny into one unified process, eliminating the perceive spatial and temporal differences between the two based on descriptive biology, now all that is left are the energetic changes, the forms being 'nodes' for the energetic 'pathways'. This perspective begs the question as to the origin of this pathway. Since it has been hypothesized that the unicell formed in reference to the Singularity (Torday, 2019), the energy vector would appear to have been the consequence of the Big Bang.

Endosymbiosis, the means by which life has evolved

Multicellular organisms appeared on earth approximately 600 million years ago. It has been hypothesized that they evolved in response to bacteria affecting pseudo-multicellular forms based on Biofilm and Quorum Sensing, but these are ad hoc structures that form to cope with the environment. In response to this existential threat to eukaryotic unicellular organisms, the latter evolved cell-cell communications mediated by soluble growth factors and their receptors in service to true multicellular properties.

The prevailing theory of cellular evolution is Symbiogenesis, the internalization of factors in the environment that presented as existential threats. Primordial cells spontaneously formed from micelles with semipermeable cell membranes that accommodated acquisition of various substances. Internal membranes then compartmentalized such substances and organized them biochemically, forming the basis for physiology (Torday and Rehan, 2012, 2017). The endogenized substances in the Cosmos are ordered by the way that stars produce the Elements during the process of stellar nucleosynthesis (Smolin, 1999), beginning on the far left of the Periodic Table as Hydrogen, progressing from left to right 'periodically' as heavier and heavier substances, with more and more protons in their nuclei (Scerri, 2006). Thus, the Elements form the hierarchic basis for the Laws of Nature, to which both the Cosmos and living organisms comply. As a consequence, our consciousness is congruent with the 'Consciousness' of the Cosmos (Torday, 2020).

'Phenotype as agent' Highlights the significance of energy flow

Epigenetic inheritance is characterized as the ability of the organism to identify specific changes in the environment, assimilate them in the egg and sperm of the ovaries and testes, respectively, and pass such information on to the offspring. Consequently, the organism has foreknowledge of environmental changes, and is able to modify its phenotype accordingly. The process by which the embryo incorporates such epigenetic data is dependent on cell-cell communications mediated by high energy phosphate 'second messengers' constrained by homeostasis. Therefore, the phenotype is acting as an 'agent' for the effective transfer of energy from one generation to the next in order to adapt to the ever-changing environment. Seen in this way, phenotypic change is a means to an end, not an end in itself as Darwinian evolution would have us think (1859).

Television sets, phenotypes and wiring

By way of introduction to this idea, we have made systematic errors in judgement based on our Explicately evolved senses (Bohm, 1980) in the past, like the earth as the center of the Solar System, the earth is flat, or spontaneous generation. So in the context of a television set, images on the screen formed by electrons interacting with the rare earth coating on the inside of the screen. And then there are the wires in the back of the TV that conduct the flow of electrons. It is there that the origin and principles involved emanate from. It is like the images on the screen as phenotypes, the electron flow being the basis for such forms and sounds.

Coherence, but to what end ?

In his masterwork, "Wholeness and the Implicate Order" (1980), David Bohm addresses how we cohere, but he does not tell us what we cohere with. Based on the precept that the totality of the Cosmos is expanding (Overbye, 2017), following the energetic vector formed by the Big Bang, coherence would be to that vector. Deviating from the vector can occur, but causes stress, which the organism may accommodate through auto-engineering (Shapiro, 2011), but that can only occur within limits. If the stress is too great, the organism will not be able to maintain its coherence, and will become extinct as a result.

Implications of focusing on energy flow as evolution

William of Occam maintained that the simplest answer was most likely the correct answer (Russell, 2000). On the other hand, Darwinian evolution gives rise to a complicated perspective on life— *"It is interesting to contemplate a tangled bank, clothed with many plants of many kinds, with birds singing on the bushes, with various insects flitting about, and with worms crawling through the damp earth, and to reflect that these elaborately constructed forms, so different from each other, and dependent upon each other in so complex a manner, have all been produced by laws acting around us. These laws, taken in the largest sense, being Growth with reproduction; Inheritance which is almost implied by reproduction; Variability from the indirect and direct action of the conditions of life, and from use and disuse; a Ratio of Increase so high as to lead to a Struggle for Life, and as a consequence to Natural Selection, entailing Divergence of Character and the Extinction of less improved forms. Thus, from the war of nature, from famine and death, the most exalted object that we are capable of conceiving, namely, the production of the higher animals, directly follows.*

There is grandeur in this view of life, with its several powers, having been originally breathed by the Creator into a few forms or into one; and that, whilst this planet has gone circling on according to the fixed law of gravity, from so simple a beginning endless forms most beautiful and most wonderful have been, and are being evolved." (Darwin, 1859)

It is proposed that when evolution is seen as energy instead of as matter that the paths it follows, as a 'forest and trees' problem, are much simpler when seen as the former than the latter. And in fact, when the criterion of evolution as 'serial pre-adaptations', or exaptations (Gould and Vrba, 1982) is applied, there are many places along the way where the connections prove impossible based on material appearance. Take the homology between the swim bladder of a fish and the lung of a mammal.

The analogy would be between the gills and the lung, both mediating gas exchange, as has been the case for decades. However, with the advent of functional genomics, it has become clear that the homology, or being of the same origin, is that of the swim bladder and lung (Torday and Rehan, 2007). Similarly, there had been a bottleneck in the pursuit of the evolution of the Central Nervous System, until Nicholas Holland pointed out that worms have their nervous systems in their skin (Holland, 2003), providing a way to connect the dots.

Moreover, there are deep cellular-molecular homologies between the skin and brain arising from the mechanism for the formation of the stratum corneum as a barrier against bacterial invasion from the outside and loss of fluid and electrolytes from the vasculature. The mechanism for depositing lipids and host-defense peptides in the skin is homologous with mechanism of secreting lung surfactant in the alveoli (Aberg et al., 2008), and myelination of neurons by Schwann Cells (Lemke, 2006). Such molecular homologies help in understanding, for example, why patients with neurodegenerative diseases like Tay Sachs, Niemann-Pick and Schizophrenia have atopic dermatitis, or Atopy, a skin lesion. Similarly, patients with Asthma also have Atopy (Custovic et al., 2013).

The genetic tie between these seemingly disparate traits is Parathyroid Hormone-related Protein (PTHrP), which links all of these phenotypes together. Importantly, the PTHrP Receptor 'duplicated' during the water-land transition, amplifying that signaling pathway in numerous tissues and organs, allowing for terrestrial life, among them being the lung, skin and brain.

Key to understanding such interrelationships is the controversy over Darwinian gradualism and Eldredge and Gould's 'punctuated equilibrium' (Gould and Eldredge, 1993). Like Bohr's explanation for the duality of light as both a particle and wave, it is a function of how you measure it (Bandyopadhyay, 2000). Seen from a molecular perspective, the same evolutionary trait occurs early on as a microscopic or sub-microscopic change, but over time it becomes visible, hence the difference in perspective.

Albert Szent-Gyorgyi (1960), a founder of modern biochemistry, said that life is an interposition between two energy levels of an electron: the ground state and the excited state. There is evidence that light can affect the cell division in paramecia (Fels, 2009), for example. And the observation that a retinal cell can detect one photon (Tinsley et al., 2016) lends credence to the role of Quantum Mechanics in biology.

Insights from evolution as energy flow

The overarching hypothesis is that the disruption of the Singularity by the Big Bang gave rise to a vectoral flow of energy that generated the Cosmos, and subsequently gave rise to life. That perspective naturally lends itself to the continuum from the physical to the organic, as has previously been hypothesized, the atom and the cell both being point sources (Torday and Miller, 2016a), both being homologues, exhibiting both deterministic and probabilistic characteristics (Torday, 2018). This reductionist perspective lends itself to reconsideration of the very nature of evolution, particularly given the re-emergence of epigenetic inheritance based on the primacy of the unicellular state (Torday, 2015)—epigenetic marks are assimilated by the egg and sperm, and subsequently incorporated into the offspring. Consequently, we now realize that we never actually leave the unicellular state, instead delegating the offspring as an 'agent' for collecting epigenetic data from the environment (Torday and Miller, 2016b). Seen in this light, evolutionary adaptation is for optimizing detection of environmental perturbations, emphasizing changes in energy flow as the characteristic to be monitored. The phenotypic changes that ensue are meant to rectify such changes in energetic flow.

The Darwinian theory of evolution has heavily influenced contemporary thought, from sociology (Wilson, 2007), to psychology (Buss, 2005), literature, and astronomy (Smolin, 1999). Yet we continue to flounder, looking for ways to formulate a 'Theory of Everything' (Weinberg, 2011) because we are the only organism that is destroying ourselves and the planet. It is proposed that we have focused on the material aspect of life, when in fact it is the energy side of $E = MC^2$ that holds the answer.

Given the above, if, for example, the literature were to focus on energy, perhaps it would be more consistent with the vector of the Big Bang.

Similarly, instead of focusing on 'supply and demand' economics would be centered on the amount of Gibbs Free Energy in the system, consistent with its relationship to physiology (Spencer-Brown, 2008). And history would be taught based on the Free Energy available to society, rather than on personalities and philosophies.

Discussion

The most compelling argument against Darwinian evolution theory is that it is illogical, founded on reasoning after the fact based on descriptive biology. Absent any other means of understanding evolution, Darwinism has prevailed, up until now. But with the resurgence of epigenetic inheritance, particularly as it applies to evolution (Torday and Rehan, 2012, 2017), there is an alternative way of considering evolution from its unicellular origins.

Historically, the major advances in our perception of our environment have countered common sense perceptions thanks to empiricism—Copernican Heliocentrism as the Sun being the center of the solar system, the Earth being round, life as spontaneous generation. Similarly, the present hypothesis that evolution is the flow of energy, not Darwin's 'tangled bank' (1859), is solely founded on experimental evidence (Torday and Rehan, 2012).

The breakthrough in our understanding of embryonic development was the discovery of soluble growth factor paracrine signaling for structure and function, mediated by second messengers as high energy phosphate compounds (Basson, 2012).

Once that mechanism was applied to phylogeny as the long-term history of speciation, it was realized that this was a solution for understanding evolution from its origins instead of its results (Torday and Rehan, 2012). Seeing the evolutionary process in the forward direction, from the unicell onward provided a causal relationship emanating from its physical origins in the Big Bang (Torday, 2019). Now, seen as a continuous process instead of as haphazard random mutations (Torday and Rehan, 2007), it became clear that we should have been focused on the serial energy exchanges, beginning with the Big Bang.

Contemporary biology and physics are at a critical phase, unable to reduce their problems to practice using their reductions to principle. In the case of biology, biomedical research is in crisis (Alberts et al., 2014), unable to bridge the gap between the gene and the phenotype. And in the case of physics, Quantum Mechanics is unable to explain everyday realities in the way that Newtonian physics has been able to. It would appear that we have reduced these complex problems to absurdity, or reductio ad absurdum. The problem seems to lie in our after-the-fact rationalization of our own beginnings and evolution. But up until now, all we had was a compendium of organisms, beginning with Linnaeus's Binomial Nomenclature, now reduced to genes. But genes do not form structures and functions, cells do. This mismatch is due to a systematic error on the part of the evolutionists, deciding to side with genetics in order to advance their knowledge, skipping over cell biology, to this day (Smocovitis, 1996).

As for the physicists, they advanced our knowledge mathematically, but now realize that there is a seeming disconnect between Quantum Mechanics and day-to-day reality (Frauchiger and Renner, 2018).

This chapter is a plea for continued scientific inquiry in an era when scientists are defaulting to spiritualism as their only seeming alternative. Take the Galileo Commission, for example, and the Non-duality of Science, or SAND group (www.scienceandnonduality.com), opting for belief instead of rational scientific method. This failure to pursue scientific quests based on hypothesis testing Popperian science is due to a lack of imagination.

References cited

Aberg, K.M., Man, M.Q., Gallo, R.L., Ganz, T., Crumrine, D., Brown, B.E., Choi, E.-H., Kim, D.-K., Schroder, J.M., Feingold, K.R. and Elias, K.R. 2008. Co-regulation and interdependence of the mammalian epidermal permeability and antimicrobial barriers. J. Invest. Dermatol. 128: 917–925.

Alberts, B., Kirschner, M.W., Tilghman, S. and Varmus, H. 2014. Rescuing US biomedical research from its systemic flaws. Proc. Natl. Acad. Sci. U.S.A. 111: 5773–5777.

Bandyopadhyay, S. 2000. Welcher weg experiments and the orthodox Bohr's complementarity principle. Phys. Lett. 276: 233–239.

Basson, M.A. 2012. Signaling in cell differentiation and morphogenesis. Cold Spring Harb. Perspect. Biol. 4: a008151.

Bohm, D. 1980. Wholeness and the Implicate Order. Routledge, London.

Buss, D.M. 2005. The Handbook of Evolutionary Psychology. Wiley, Hoboken.

Crick, F. 1970. Central dogma of molecular biology. Nature 227: 561–563.

Custovic, A., Lazic, N. and Simpson, A. 2013. Pediatric asthma and development of atopy. Curr. Opin. Allergy Clin. Immunol. 13: 173–180.

Darwin, C. 1859. On the Origin of Species. John Murray, London.

Deamer, D. 2017. The role of lipid membranes in life's origin. Life 7: 5.

Fels, D. 2009. Cellular communication through light. PLoS One 4(4): e5086.

Frauchiger, D. and Renner, R. 2018. Quantum theory cannot consistently describe the use of itself. Nat. Commun. 9: 3711.

Frappier, M., Meynell, L. and Brown, J.R. 2012. Thought Experiments in Science, Philosophy, and the Arts. Routledge, London.

Gould, S.J. and Vrba, E.S. 1982. Exaptation—a missing term in the science of form. Paleobiology 1: 4–15.

Gould, S.J. and Eldredge, N. 1993. Punctuated equilibrium comes of age. Nature 366: 223–227.

Hawking, S. 1988. A Brief History of Time. Bantam, New York.

Henry, J. 2001. Moving Heaven and Earth: Copernicus and the Solar System. Icon, Cambridge.

Holland, N.D. 2003. Early central nervous system evolution: an era of skin brains? Nat. Rev. Neurosci. 4: 617–627.

Lemke, G. 2006. Neuregulin-1 and myelination. Sci. STKE 200: pe11.

Margulis, L. and Chapman, M.J. 2009. Kingdoms and Domains: an Illustrated Guide to the Phyla of Life on Earth. Academic Press, Amsterdam.

Overbye, D. 2017. Cosmos Controversy: the Universe Is Expanding, but How Fast? The New York Times, New York.

Penzias, A.A. and Wilson, R.W. 1965. A measurement of excess antenna temperature at 4080 Mc/s. Astrophys. J. Lett. 142: 419–421.

Russell, B. 2000. History of Western Philosophy. Allen & Unwin, Crow's Nest.

Scerri, E. 2006. The Periodic Table. Oxford University Press, Oxford.

Shapiro, J.A. 2011. Evolution: A View from the 21st Century. FT Press, Hoboken.

Smocovitis, V.B. 1996. Unifying Biology. Princeton University Press, Princeton.

Smolin, L. 1999. The Life of the Cosmos. Oxford University Press, Oxford.

Spencer-Brown, G. 2008. Laws of Form. Bohmeier, Leipzig.

Szent-Gyorgi, A. 1960. Introduction to a Submolecular Biology. Academic Press, New York.

Tinsley, J.N., Molodtsov, M.I., Prevedel, R., Wartmann, D., Espigule-Pons, J., Lauwers, M. and Vaziri, A. 2016. Direct detection of a single photon by humans. Nat. Commun. 7: 12172.

Torday, J.S. and Rehan, V.K. 2007. The evolutionary continuum from lung development to homeostasis and repair. Am. J. Physiol. Lung Cell Mol. Physiol. 292: L608–L611.

Torday, J.S. and Rehan, V.K. 2009. Lung evolution as a cipher for physiology. Physiol. Genom. 38: 1–6.

Torday, J.S. and Rehan, V.K. 2012. Evolutionary Biology, Cell-Cell Communication and Complex Disease. Wiley, Hoboken.

Torday, J.S. 2015. The Cell as the Mechanistic Basis for Evolution, vol. 7. Wiley Interdiscip. Rev. Syst. Biol. Med., pp. 275–284.

Torday, J.S. and Miller, W.B. 2016a. The unicellular state as a point source in a Quantum biological system. Biology 5: 25.

Torday, J.S. and Miller, W.B. 2016b. Phenotype as agent for epigenetic inheritance. Biology 5: 30.

Torday, J.S. and Rehan, V.K. 2017. Evolution, the Logic of Biology. Wiley, Hoboken.

Torday, J.S. 2018. Quantum Mechanics predicts evolutionary biology. Prog. Biophys. Mol. Biol. 135: 11–15.

Torday, J.S. 2019. The Singularity of nature. Prog. Biophys. Mol. Biol. 142: 23–31.

Torday, J.S. 2020. Consciousness, redux. Med. Hypotheses 140: 109674.

Torday, J.S. 2021. Cellular evolution as the flow of energy. Progr. Biophys. Mol. Biol. 167: 147–151.

Weinberg, S. 2011. Dreams of a Final Theory: the Scientist's Search for the Ultimate Laws of Nature. Knopf Doubleday Publishing Group, New York.

Whitehead, A.N. 1929. Process and Reality. Macmillan, New York.

Wilson, D.S. 2007. Evolution for Everyone: How Darwin's Theory Can Change the Way We Think about Our Lives. McHenry, Delta.

Chapter 3
It's Epigenetics 'All the Way Down'

The results of an experiment published in 2003 demonstrated that when differentiated cells are exposed to zero gravity they lose their phenotypic identity (Torday, 2003). This result speaks volumes about epigenetics, given that the genome was still intact, but the loss of cell membrane integrity led to loss of the phenotype. In a series of experiments, the significance of the cell membrane in evolution was highlighted by the fact that cell-cell communication through soluble growth factors and their receptors was key to understanding the role of memory in the process of evolution. In point of fact, the ability to trace the evolution of complex physiology from its end to its beginning in the unicell (Torday and Rehan, 2012) is experimental evidence that evolution is not due to random mutations, it is shaped by its environment, initiated by the force of gravity.

The value of this perspective is underscored by the post-diction that Quantum Mechanics underlies the evolutionary process. In retrospect, the reverse-engineering of lung evolution had been accomplished by tracing the sequential emergences involved in its history (Torday and Rehan, 2017), all the way back to the step in which cholesterol was synthesized in association with the rise in atmospheric oxygen. Konrad Bloch had hypothesized that this occurred because it takes 11 atoms of oxygen to synthesize one molecule of cholesterol, referring to the latter as a 'molecular fossil'. But he was reasoning after the fact. Considering this event as an 'exaptation' (Gould and Vrba, 1982), what pre-adaptations were available? There is experimental evidence in the forward direction for snowball-like asteroids providing lipids and water to the nascent Earth, and lipids have been shown to protect the cell from oxidative damage (Torday et al., 2001). Therefore, it was hypothesized that there was selection pressure for lipids like cholesterol to protect the burgeoning unicell against oxygen, particularly since lipids are amphiphiles, having charged poles. When the Earth's gravity pulled downward on those lipids, they aligned vertically to the air-water interface, packing together, their negatively-charged hydrophilic poles facing downward into the water. When the aggregate negative charge had reached a critical level it was able to 'neutralize' the Van der Waals force for surface tension, freeing the lipids to spontaneously form micelles, or prototypical cells. The space within those cells allowed for the emergence of the First Principles of Physiology—negative entropy, chemiosmosis and homeostasis (Torday and Rehan, 2009). The force of gravity impinging on the curved surface of the micelle generated energy (Einstein, 1961) for Quantum Entanglement of the particles that had entered the cell through the semi-permeable membrane of the micelle, providing the first empiric evidence for the role of Quantum Mechanics in the evolution of the cell (Torday, 2021), complete with the local effect of gravity referencing the non-local gravitational force of the Cosmos. It is that local/non-local characteristic of the cell that forms the basis for the fractal nature of the cell. The cell acts as a cipher for the force of gravity, providing the connection between life and non-life.

The above was the means for the 'discovery' of step-wise evolution of life from the interaction between the local force of gravity referencing the non-local force of gravity in the Cosmos as one of the four forces produced by the Big Bang, tracing the process from its inception based on experimental evidence.

In contrast to the above, those feigning empiricism such as Sheldrake and Levin completely ignore the 13,000 peer-reviewed articles on cell-cell signaling that undergirds both ontogeny and phylogeny, allowing for evidence of Haeckel's "Ontogeny Recapitulates Phylogeny", both processes being mediated by the same communication mechanism. Not to mention that two Nobel Prize winners who won by identifying Epidermal Growth Factor (Cohen, 1987) and Nerve Growth Factor (Cowan, 2001).

The reduction of ontogeny and phylogeny to cell-cell communications mediated by soluble growth factors and their cognate receptors allowed for their superposition, mimicking a wave collapse in Quantum Mechanics, leading to the realization that life is primarily energy flow, not material traits. This is particularly true of embryogenesis, which tends to focus on the morphologic changes that occur over the course of development from the zygote to the offspring, when in fact when seen from the perspective of cell-cell communication, it is a series of high energy phosphate exchanges caused by the growth factor/growth factor receptor interactions producing either cyclic AMP or inositol phosphates that ultimately affect the readout of DNA in the cell nucleus.

That process culminates in physiologic homeostasis in the offspring, also acting to regulate the flow of energy through the organism. Seen in that light, the phenotype is not merely an inventory of biologic traits, it is an 'agent' for detecting change in the environment, or epigenetic marks, which are assimilated by the egg and sperm, allowing for adaptation. The inherent truncation of the life cycle from its conventional stages—newborn, adolescent, adult—to the free flow of energy fits better with A.N. Whitehead's Process Philosophy than with Darwin's "Descent with Modification", the former emphasizing energy flow, the latter emphasizing material being. Based on Occam's Razor the former is much simpler than the latter. Consider Darwin's 'Tangled Bank', "It is interesting to contemplate a tangled bank, clothed with many plants of many kinds, with birds singing on the bushes, with various insects flitting about, and with worms crawling through the damp earth, and to reflect that these elaborately constructed forms, so different from each other, and dependent upon each other in so complex a manner, have all been produced by laws acting around us." (1859).

Seen in this way, evolution becomes predictive rather than descriptive. For example, it was serendipitously discovered several decades ago that Parathyroid Hormone-related Protein (PTHrP) is necessary for the formation of alveoli in the developing lung (Rubin et al., 2004). The subsequent discovery that PTHrP is produced by lung fibroblasts, stimulating lung surfactant synthesis by neighboring alveolar type 2 cells provided an explanation for 'how and why' it was needed for lung development.

That relationship led to the demonstration that PTHrP expression predicts the risk of Bronchopulmonary Dysplasia (Rehan and Torday, 2006), or chronic lung development that bridges the developmental gap through reverse evolution. The net result is that the surviving offspring can pass its genes on to the next generation. This perspective is 180 degrees out of sync with conventional medicine, which sees BPD as a disease, not as an adaptation. It has been used to encourage a paradigm shift away from disease, towards evolutionary medicine as the basis for Preventive Medicine by recognizing the above relationships in order to pre-empt the person as a patient.

In fact, the above brings us full circle since the serendipitous discovery that the hormone cortisol accelerates normal lung development via soluble growth factors and their receptors as the impetus for considering evolution in light of cellular cooperativity. Prior to that, prematurely born infants were treated PRN, or "as needed", whereas with the inception of the use of corticoids to accelerate lung development *in utero*, the fetus became a patient, giving rise to the medical discipline of Neonatology.

Perhaps this is a lesson in the purpose of life. Evolution has enabled Man to imagine what it would be like to do 'x', acting in lieu of physics to provide solutions. That imagination evolved from the unicell as a mobius strip (Torday, 2021), the cell being able to imagine what it was before it was a cell, as lipid molecules floating in the primordial ocean. Aided by bipedal locomotion, endothermy, toolmaking/language, we have figured out how to transcend our Earthly existence. Now we need to correct the mistakes produced by a lack of such insight and understanding in order to advance.

References cited

Cohen, S. 1987. Epidermal growth factor. *In Vitro* Cell Dev. Biol. 23: 239–246.

Cowan, W.M. 2001. Viktor Hamburger and Rita Levi-Montalcini: the path to the discovery of nerve growth factor. Annu. Rev. Neurosci. 24: 551–600.

Darwin, C. 1859. On the Origin of Species. John Murray, London.

Einstein, A. 1961. Relativity. The Special and General Theory. Crown Publishing, New York.

Gould, S.J. and Vrba, E.S. 1982. Exaptation—a missing term in the science of form. Paleobiology 8: 4–15.

Rehan, V.K. and Torday, J.S. 2006. Lower parathyroid hormone-related protein content of tracheal aspirates in very low birth weight infants who develop bronchopulmonary dysplasia. Pediatr. Res. 60: 216–220.

Rubin, L.P., Kovacs, C.S., De Paepe, M.E., Tsai, S.W., Torday, J.S. and Kronenberg, H.M. 2004. Arrested pulmonary alveolar cytodifferentiation and defective surfactant synthesis in mice missing the gene for parathyroid hormone-related protein. Dev. Dyn. 230: 278–289.

Torday, J.S., Torday, D.P., Gutnick, J., Qin, J. and Rehan, V. 2001. Biologic role of fetal lung fibroblast triglycerides as antioxidants. Pediatr. Res. 49: 843–849.

Torday, J.S. 2003. Parathyroid hormone-related protein is a gravisensor in lung and bone cell biology. Adv. Space Res. 32: 1569–1576.

Torday, J.S. and Rehan, V.K. 2009. Lung evolution as a cipher for physiology. Physiol. Genomics 38: 1–6.

Torday, J.S. and Rehan, V.K. 2012. Evolutionary Biology, Cell-Cell Communication and Complex Disease. Wiley, Hoboken.

Torday, J.S. and Rehan, V.K. 2017. Evolution, The Logic of Biology. Wiley, Hoboken.

Torday, J.S. 2021. Life is a mobius strip. Prog. Biophys. Mol. Biol. 167: 41–45.

Chapter 4

Ontogeny, Phylogeny and Cellular Energy Flows for Evolution

Introduction

The case has previously been made for the 'Singularity of Nature' (Torday, 2019) based on homologies between the inorganic and the organic. But a causal explanation for those homologies was not provided. The following is a hypothetical way to understand how and why physiology emerged from physics, allowing for falsifiability (Popper, 1952).

Briefly, gravity is necessary for the formation (Classen and Spooner, 1996) and maintenance of life (Torday, 2003), raising the question as to how and why it has facilitated ontogeny and phylogeny. In an earlier publication, the duplication of specific receptor genes was structurally and functionally linked to the transition from water to land some 500 million years ago. But it leaves open the question as to which organ-bone, lung, skin-initiated the step-wise progression in the adaptation to land? Known to have been attempted on at least 5 separate occasions (Clack, 2012). The key to understanding this lies in the context the organism finds itself in, in this case the depletion of oxygen from the water due to the 'greenhouse effect', the solubility of oxygen in water being a function of temperature, in combination with the increased effective force of gravity on land (versus water) (Romer, 1949). Insight to this question was found in the mechanism of action of nicotine on asthma, stimulating the N-Acetylcholine Receptor in the smooth muscle of the upper airway (Sakurai et al., 2011). Asthma is deleterious for lung function, but in the skeleton this pathway facilitates the remodeling of bone (Mandl et al., 2016). In other words the selection pressure for limb evolution in

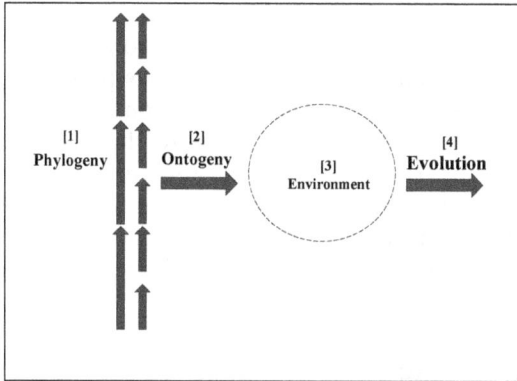

Figure. Cell-Molecular Evolution in Environmental 'context'. The on-going processes of phylogeny [1] and ontogeny [2] (= vertical arrows) respond to an ever-changing environment in order to adapt [3]. The net result is evolution [4].

transition from water to land had a pleiotropic effect on the swim bladder pneumatic duct (Finney et al., 2006), facilitating the evolution of the lung.

In this context, the cellular-molecular approach to ontogeny and phylogeny is predictive, the N-Acetylcholine Receptor's effect on bone remodeling pre-empting that on the smooth muscle of the swim bladder pneumatic duct during the transition from water to land.

Therefore, the spatio-temporal nature of cellular evolution can be exploited in order to understand how and why tissues and organs have evolved. This somewhat convoluted way of understanding the interrelationship between the organism and its environment across space-time should be kept in mind as it has evolved from the physical to the organic. In retrospect, the interrelationship of the water-land transition to both gravity and oxygen should not have come as a surprise, yet it is not the way we conventionally think because it is diachronic, across space-time, yet that is the way we must think about evolution in order to subordinate the material changes in favor of the way that the process facilitates the flow of energy. It is to that principle that the following is dedicated.

Stellar nucleosynthesis, or starlight, produces the First 36 Elements

Stellar nucleosynthesis is the process by which stars generate light through nuclear reactions of previously existing nuclei or nucleons, matter in the form of the Elements being byproducts. Initially, this process produced hydrogen, helium and lithium, followed subsequently by another 33 Elements, the last of which in the series was krypton, atomic number 36. The relationships of the Elements are documented in Mendeleev's

Periodic Table of Elements (PTE), beginning with hydrogen, in a sequence based on their atomic mass, or the number of protons in their nuclei. Each Element is characterized in two ways, (1) by its physical characteristics, as is the case for Alchemy, and (2) by its atomic mass, referencing its origin from stellar nucleosynthesis. Therefore, the PTE is simultaneously synchronic, reflecting its characteristics in real-time, as well as diachronic, based on its characteristics across space-time. And because Quantum Entanglement forms the juncture with Symbiogenesis, the local/non-local fractal property of the cell forms the link between the inanimate and the animate.

The Elements of the periodic table form a 'logic' that is the basis for physiology

Frescura and Hiley (1984) state that if you juxtapose two harmonic processes at right angles to one another, they generate cycles. Such cyclicity is expressed when organisms assimilate factors in the environment that pose existential threats, referred to by Lynn Margulis Sagan as Symbiogenesis (1967). Thus, the Elements have been assimilated over the course of evolution, being made useful within us as physiologic traits (Torday and Rehan, 2012). As a result, they naturally reference the Elements in the Cosmos, and therefore our physiology ascribes to the same logic as the Cosmos (Torday and Rehan, 2017).

Cell-cell communication as the basis for physiologic evolution

Over the course of vertebrate evolution, traits have evolved in order to adapt to an ever-changing environment as a consequence of the expanding Universe (Hawking, 2011). In order to evolve effectively and keep up with environmental change, the organism must be able to reference its 'history'. That is accomplished through cell-cell communications mediated by soluble growth factors and their cognate receptors (Torday and Rehan, 2012). Confronted by a novel existential threat, the organism is able to assimilate such factors and relate them to its genomic history in the germ cells (egg and sperm), referred to as Epigenetic Inheritance. Over the course of meiosis, or reduction division, the gametes are able to discern whether any given epigenetic 'mark' should or should not be incorporated into the genome through methylation or ubiquitination of specific sequences in the DNA code (King and Skinner, 2020). As a result, the offspring will express those segments of DNA code as adaptive modifications. If such epigenetic inheritances persist due to unmitigated factors in the environment, they will result in change in the DNA code as evolutionarily adaptive changes.

Terminal addition codifies the logic of the cosmos

The phenomenon of Terminal Addition in biology is well recognized, newly acquired evolutionary traits being added on to the end of a series of evolved traits, rather than in the middle or at the beginning. The reason for this is explained in "Terminal Addition in a Cellular World" (Torday and Miller, 2018) as follows:

> "Recent advances in our understanding of evolutionary development permit a reframed appraisal of Terminal Addition as a continuous historical process of cellular-environmental complementarity. Within this frame of reference, evolutionary terminal additions can be identified as environmental induction of episodic adjustments to cell-cell signaling patterns that yield the cellular-molecular pathways that lead to differing developmental forms. Phenotypes derive, thereby, through cellular mutualistic/competitive niche constructions in reciprocating responsiveness to environmental stresses and epigenetic impacts. In such terms, Terminal Addition flows according to a logic of cellular needs confronting environmental challenges over space-time. A reconciliation of evolutionary development and Terminal Addition can be achieved through a combined focus on cell-cell signaling, molecular phylogenies, and a broader understanding of epigenetic phenomena among eukaryotic organisms. When understood in this manner, Terminal Addition has an important role in evolutionary development, and chronic disease might be considered as a form of 'reverse evolution' of the self-same processes."

When Terminal Addition is considered in the context of symbiogenesis, or the endogenization and assimilation of environmental factors, the 'logic' of the Cosmos dictates both the inanimate and the animate.

The First 36 Elements of the Periodic Table form the basis for the logic of physiology….. Iodine, #53, is the 'exception to the rule' which demonstrates how/why physiology has learned to emulate the cosmos

Stellar nucleosynthesis is the process by which stars generate light. In so doing, they produced the first 36 Elements of the Periodic Table, from hydrogen to krypton, many of which have been assimilated for physiologic function. An exception that proves the rule is iodine, Element 53, which was used by the thyroid gland to synthesize thyroxin by chemically bonding iodine with tyrosine.

The phylogenies of the endostyle and non-follicular thyroid tissues of vertebrates and invertebrates has been documented (Dickhoff and Darling,

1983), emphasizing the roles of exogenous compounds and peripheral mechanisms in regulating thyroid status, and influencing the course of the phylogenetic development of the vertebrate thyroid gland. Central to understanding the adaptive role of thyroid physiology is the utilization of iodine, for which there are numerous non-chordate invertebrates that have the ability to bind iodine. Free iodo-compounds, either eaten or absorbed, and even iodothyronines from plants and microorganisms, can be metabolized in the gut aided by resident bacteria. The intestine is the primary site for mono-deiodination from thyroxin to triiodo thyronine, which is also true for ascidians. The endostyle is only found in marine invertebrates that have a notochord, and in freshwater larval lampreys. This specialized region of the alimentary tract may have evolved due to positive selection pressure for filter feeding. Like many invertebrate tissues, it had iodine-binding capacity, which developed secondarily. The iodine-binding capacity of the endostyle in existing ascidians and larval lampreys is well documented. The retention of the endostyle over the course of the evolutionary history of lampreys is indicative of its importance in the filter-feeding apparatus, suggesting that it may very well have appeared during the pelagic marine phase of lamprey history.

The vertebrate follicular thyroid gland evolved due to strong positive selection pressure for a gland that had the capacity for thyroid hormone and iodine storage when ancient chordates transitioned from an iodine-rich marine environment to an iodine-poor freshwater habitat. In this context, the presence of an endostyle in extant larval lampreys reflects their ancient marine origins, the adult follicular thyroid gland appearing after they moved toward freshwater. Since the follicular thyroid gland of lampreys only appears after the transformation from the endostyle, there was selection pressure generated by the freshwater environment for metamorphosis in the ontogeny of lampreys. Consequently, this metamorphosis characterized the transformation of the endostyle into the follicular thyroid gland.

The ontogeny of the lamprey thyroid gland in combination with the freshwater habitat of lampreys as a secondarily acquired niche implies that metamorphosis might not have originated as a developmental strategy. Rather, it may have occurred when lampreys moved into freshwater during their evolutionary history. Alternatively, the follicular thyroid gland of the marine environment was the primary reason why lampreys could inhabit freshwater. The endostyle of early lampreys was not a prerequisite for living in a marine environment, since modern-day larvae with endostyles cannot tolerate even dilute seawater. Pre-metamorphic lampreys have lost the ability of the hypothetical ancestral larval form to inhabit a marine environment. It is only after metamorphosis that juveniles of some species are able to tolerate seawater.

Larval and reproductive phases of the lamprey lifecycle occur exclusively in freshwater. The capacity of juveniles of some species to osmo-regulate in seawater could also indicate that was secondarily derived following the evolution of metamorphosis in the ontogeny of lampreys. Existing parasitic lampreys of the most ancient lineage (e.g., Ichthyomyzonunicupis) exclusively live in freshwater. An explanation for the evolution of marine osmoregulation may secondarily have been derived following the primary derivation of metamorphosis based on contemporary views referred to as *developmental integration*.

Evolutionary phylogenetic and ontogenetic vertical integration of the thyroid gland

The increased bacterial load due to facilitation of feeding by the endostyle may have stimulated the cyclic Adenosine Mono-Phosphate-dependent Protein Kinase A, or PKA pathway since bacteria produce endotoxin, a potent PKA stimulant. This molecular cascade may have evolved into regulation of the thyroid by thyroid stimulating hormone (TSH), since TSH acts on the thyroid via the cyclic Adenosine Mono-Phosphate, or cAMP-dependent PKA signaling pathway. This mechanism hypothetically generated novelties such as the thyroid, lung, and pituitary, all of which are developmentally induced by the PKA-sensitive Nkx2.1/TTF-1 pathway. Brain–lung–thyroid syndrome, in which infants with Nkx2.1/TTF-1 mutations develop hypotonia, hypothyroidism, and respiratory distress syndrome, or surfactant deficiency disease as further evidence for the coevolution of the lung, thyroid, and pituitary.

During development the thyroid buds off of the foregut beginning on day 8.5 in the mouse, one day before the lung and pituitary, implying that the thyroid may have been a molecular prototype of the lung over the course of evolution, providing a testable and refutable hypothesis. The thyroid allowed molecular iodine in the environment to be bioavailable by binding it to the essential amino acid threonine to generate thyroid hormone. The lung rendered oxygen bioavailable by inducing fat cell–like lipofibroblasts containing neutral lipids, acting as cytoprotectants (Torday et al., 2001), which then stimulated surfactant production by producing leptin (Torday et al., 2002), placing increased selection pressure on the blood–gas barrier, rendering the alveoli more compliant. This may have created further positive selection pressure for the metabolic system by utilizing the rising oxygen levels in the environment, placing further selection pressure on the alveoli, giving rise to the stretch-regulated surfactant system mediated by Parathyroid Hormone-related Protein and leptin. Subsequent selection pressure on the cardiopulmonary system may have facilitated liver evolution, since the progressively increasing size of the heart may have induced precocious liver development, fostering increased glucose regulation. The brain acts like a glucose sink, and there is experimental evidence that increasing glucose during pregnancy increases the size of the developing brain (Saintonge and Coté, 1984). Further evolution of the brain, specifically that of the pituitary, would have served to propel the evolution of complex physiologic systems.

Gravity is necessary for life as the basis for its latticework

Beyond the mechanism of Terminal Addition, the force of gravity has formed the basis for both planets and cells alike. As one of the four forces produced by the Big Bang, gravity has been found to underlie the formation of planetary spheres (Wurm and Teiser, 2021) and cellular spheres alike (Classen and Spooner, 1996). Resistance to gravity as the 'equal and opposite force' as Newton's Third Law of Motion likely produced life, given that in the virtual absence of gravity differentiated cells lose their phenotypes (Torday, 2003), and in the case of yeast they cannot polarize or bud, i.e., reproduce (Purevdorj-Gage et al., 2006).

Synchronicity, a product of latticework

It was Carl Jung who hypothesized what he called 'synchronicities', things that happened concurrently (Sacco, 2019) due to different causes as a consequence of the interrelationship between the force of gravity impinging on the curvature of the protocell producing energy, maintaining Quantum Entanglement (QE) and non-localization of gravity. Consequently, spiritualism and science both emerge from the same effect of gravity.

Phantom limb, terminal addition and latticework

Phantom Limb is a well-recognized phenomenon- sensing feeling in a severed limb. It makes no sense to expend physical and psychic energy on a non-existent limb, but thought of in the context of Terminal Addition (see above), the cell-cell signaling upstream from the severed limb must be maintained to avoid disuse atrophy. This mechanism is consistent with the latticework of our physiology.

Homeobox gene co-linearity as added proof of principle

The homeobox genes determine the structure of the organism. They align along the chromosome in the same way that they do along the 'head-tail axis'. The Hox gene cluster is hypothesized to have arisen due to tandem duplications from a single prototypical Hox gene (Gaunt, 2015). Subsequent duplications likely occurred due to unequal crossing-over between homologous chromosomes during meiosis. Each duplication is characterized by two similar DNA fragments lying adjacent to one another in the same orientation along the same chromosome. This mechanism for forming the first Hox cluster potentially explains why all vertebrate Hox genes are in the same transcriptional orientation.

Quantum mechanics and latticework; quantum chromodynamics theory

Working backwards from contemporary complicated physiology by exploiting cell-cell communication leads universally to the unicellular state as a function of Symbiogenesis. Initially, the lung was used as a prototype for the reverse-engineering of cellular evolution (Torday and Rehan, 2012), identifying key emergences over the course of vertebrate evolution. Beginning with the evolution of surfactant from its origin in the fish swim bladder (Daniels et al., 2004), where it acts to prevent the walls of the bladder from adhering to one another. Prior to that, lipids in peroxisomes were utilized to buffer the leaking of calcium into the cytoplasm (de Duve, 1969). And even earlier and more fundamental was the emergence of cholesterol, which facilitated oxygenation, metabolism and locomotion when inserted into the cell membrane (Bloch, 1992). Other evolved structures such as the kidney glomerulus, calcification of bone and the thyroid have been traced from their present forms and functions back to their unicellular status (Torday and Rehan, 2017).

But that raises the question as to where symbiogenesis emerged from as a reference to both internal and external environmental conditions? Recalling that Einstein's Field Theory dictates that when gravity impinges on a curved surface it produces energy, the question arose as to what that energy might have been used for biologically? QE captures particles that pass through the semi-permeable membrane of the cell, so that such energy would have been used to stabilize QE, referencing non-local gravity in the Cosmos. At its origin, that process would have initiated the latticework of life, mediating the assimilation of factors in the environment. Therefore, both that which was inside the cell and that which was in the Cosmos was under the auspices of the same Laws of Nature.

Roger Penrose hypothesizes that a small amount of Quantum Space from the prior universe initiates the subsequent universe (https://www.youtube.com/watch?v=DpPFn0qzYT0), recapitulating the Quantum Entanglement that fosters Symbiogenesis (Sagan, 1967). In this vein, it is noteworthy that mathematics and physics reduce to 'nothing' (Rowlands, 2014), whereas the reduction of physiology culminates in the atoms and photons of the Quantum realm. That difference would suggest that math and physics are epiphenomena of life, like Husserl's explanation for the 'Origin of Geometry' being derived from the external environment (Husserl, 1939).

Using the 'Reverse-Evolution' approach for emergence led from symbiogenesis to quantum entanglement/ non-localization...... But where did quantum mechanics evolve from? Particle physics, including quantum chromodynamic theory as the common latticework of the cosmos

If symbiogenesis is a consequence of QE/non-localization, what gave rise to QE? Prior to the existence of the protocell there were only elementary particles in the Cosmos. Perhaps they can be accounted for by Lattice Quantum Chromodynamics such as gluons-quarks-anti-quarks.

The 'Gyrations' of hominin evolution from ambiguity to logic

Many twists and turns have occurred during the history of vertebrate evolution, beginning with the highly competitive fertilization of the egg by millions of spermatozoa. As a consequence of such competition, the ratio of male fertilized eggs to females is 4:1 in humans, referred to as the primary sex ratio; however, at birth the ratio of males to females is roughly 1:1, referred to as the secondary sex ratio. There is apparently intense positive selection pressure for specific males, hypothetically related to the sociologic phenomenon called 'Cads and Dads', or the tendency of some males to 'hit and run', while others bond with their mates for life (Durante et al., 2012). The question is what is the underlying mechanism that would explain this selection process? It is hypothetically possible that it is the consequence of excess production of Human Chorionic Gonadotropin (HCG) production by the placenta (Walton et al., 1999), stimulating fetal gonadal androgen production, inhibiting the amount of progesterone produced by the maternal ovary to maintain the pregnancy. Too much testosterone would inhibit the production of progesterone, causing differential stillbirth among males during the second trimester (McMillen, 1979).

During the course of embryologic development, the conceptus twists and turns, reproducing its phylogenetic 'history'. Wolpert has told us that the most important thing we will do in our lives is gastrulate (Hopwood, 2022). Gastrulation is the stage in embryogenesis when the mesoderm is introduced between the endoderm and ectoderm, providing for the robust plasticity of the embryo to adapt to ever-changing environmental conditions.

Perhaps the most radical adaptive change by the vertebrate conceptus was in becoming a deuterostome, developing from anus to mouth, in the opposite direction to the force of gravity. That reversal in direction was structurally and functionally reinforced by the vagus, the largest autonomic nerve, extending from the posterior level around where the adrenals are located atop the kidneys, to the gut, to the heart, to the face, to the brain, including the optic chiasm, requiring the flipping of visual images from left to right, and right to left.

It is hypothesized that it is this active effort over the course of vertebrate evolution to defy the force of gravity that ultimately gave rise to the lateralized left-right brain (McGilchrist, 2019), beginning with Quantum Mechanics at the origin of life, fostering QE and non-localization as the basis for Symbiogenesis at the inception of evolution. Indeed, QE provides the conduit from non-local Cosmic Consciousness (Bucke, 1901) to local consciousness, due to classic physics causing the interaction between the organism and its environment.

As has been related elsewhere, warm-bloodedness indirectly provided the metabolic energy for standing on two legs, or bipedalism (Lambertz et al., 2015), allowing the forelimbs to evolve for toolmaking, including language (Torday, 2021a). This process of moving in a direction away from the force of gravity, perhaps fostering deuterostomy, gave rise to our over-sized central nervous system, i.e., brain.

The history of these twists and turns is recorded in our highly evolved cerebrum, characterized as lateralized left-right brain. The specific characteristics exhibited by each have evolved in order for the brain to discern Quantum Mechanical traits deeply embedded in our psyches.

Mathematical "knots" bring us full circle' from cosmos to physiology and back

Mathematical knots form by twisting a circle into different configurations. The test for whether a knot is a 'true knot' is whether it can be unknotted to reform as a circle. In doing so, the unknotting behaves as if there were 'springs' embedded in them (Kauffman, 1994). This is very similar to what occurs during embryologic development, the zygote 'knotting' itself to generate the different intermediate forms of the offspring, stabilized by intracellular matrix. During injury-repair or aging, the matrix between cells breaks down in order for the cells to communicate again.

Cell membrane as a functional mobius strip, giving rise to imagination

The way that complicated physiology was tracked back to the unicell was by following the emergent properties along the way mediated by cell-cell communications (Torday and Rehan, 2012). At each of these existential junctures, specific pre-adaptations were reused to cope with specific existential threats in the environment. Perhaps the earliest of these emergences was brought on by the rise in carbon dioxide in the atmosphere produced by plant life, causing a 'greenhouse effect' (Romer, 1949). In response, animal cells synthesized cholesterol, a natural antioxidant, in order to protect themselves against the corrosive effects of oxygen in the environment. The appearance of cholesterol in the cell membrane had multiple effects, thinning the membrane for increased oxygen uptake, facilitating metabolism, and increasing cytoplasmic streaming that promoted locomotion. The subsequent symbiogenic (Sagan, 1967) steps mediated the evolution of physiologic traits, begging the question as to where that mechanism evolved from? When gravity impinges on a curved surface like that of the micelle, it produces energy (Einstein, 1961) that would have facilitated the QE of particles within the prototypical cell, forming what is conventionally referred to as physiology. QE as the basis for Symbiogenesis is a property of Quantum Mechanics, which is the basis for the evolution of life (Torday, 2021b). This process was made possible by the semipermeable cell membrane, and the emergent cell 'remembers' its history as individual lipid molecules, the cell membrane behaving as a functional mobius strip, having no inside or outside. That conduit from the inside of the cell to the Cosmos is the source of our imagination.

Laws of form predicts the mechanism of cell-molecular evolution

George Spencer-Brown published his "Laws of Form" in 1969 to great praise for the mathematics and philosophy. His perspective on "Marks" as distinctions in the void is very much the same as the idea that cells formed spontaneously in the primordial ocean from lipid molecules immersed in water. In both cases a space was formed that became filled, generating a consciousness. In the case of LoF it is an abstraction, whereas in the case of the cell, a mechanism has been provided by which Symbiogenesis caused the filling of the cell in order to cope with existential threats. In the aggregate, that mechanism led to the formation of what we refer to commonly as physiology. Thus, the cell-cell communications that constitute our physiology are in the aggregate what we think of as consciousness (Torday, 2020).

Conclusions

The method of thinking about organismal change in context ontogenetically, phylogenetically and evolutionarily has previously been proposed as a way of practicing true 'preventive medicine' (Torday and Rehan, 2009).

For example, given the discovery that nicotine induces asthma due to up-regulation of the N-Acetylcholine Receptor as it relates to the evolution of the skeleton during the water-land transition would have predicted the pleiotropic effect of nicotine on osteoporosis (Broulik et al., 2007). Such interrelationships are currently not thought of causally because we only think of physiology and pathophysiology synchronically, whereas the role of nicotine in stimulating N-Acetylcholine Receptors is diachronic, transcending space-time as the energy flow that is being sustained evolutionarily. That interrelationship is underscored by the heterogeneity of the N-Acetylcholine Receptors, nicotine-induced asthma specifically affecting the 3α and 7α isoforms of the receptor in the smooth muscle of the trachea (Sakurai et al., 2011).

References cited

Bloch, K. 1992. Sterol molecule: structure, biosynthesis, and function. Steroids 57: 378–383.

Broulik, P.D., Rosenkrancová, J., Růzicka, P., Sedlácek, R. and Kurcová, I. 2007. The effect of chronic nicotine administration on bone mineral content and bone strength in normal and castrated male rats. Horm. Metab. Res. 39: 20–24.

Bucke, M. 1901. Cosmic Consciousness. E.P. Dutton, Boston.

Claassen, D.E. and Spooner, B.S. 1996. Liposome formation in microgravity. Adv. Space Res. 17: 151–160.

Clack, J.A. 2012. Gaining Ground. Indiana University Press, Bloomington.

Daniels, C.B., Orgeig, S., Sullivan, L.C., Ling, N., Bennett, M.B., Schürch, S., Val, A.L. and Brauner, C.J. 2004. The origin and evolution of the surfactant system in fish: insights into the evolution of lungs and swim bladders. Physiol. Biochem. Zool. 77: 732–749.

De Duve, C. 1969. Evolution of the peroxisome. Ann. N. Y. Acad. Sci. 168: 369–381.

Dickhoff, W.W. and Darling, D.S. 1983. Evolution of Thyroid Function and Its Control in Lower Vertebrates. Am. Zool. 23: 697–707.

Durante, K.M., Eastwick, P.W., Finkel, E.J., Gangestad, S.W. and Simpson, J.A. 2012. Pair-bonded relationships and romantic alternatives: toward an integration of evolutionary and relationship science perspectives. Adv. Exp. Soc. Psych. 53: 1–74.

Einstein, A. 1961. Relativity. The Special and General Theory. Crown Publishing, New York.

Finney, J.L., Robertson, G.N., McGee, C.A., Smith, F.M. and Croll, R.P. 2006. Structure and autonomic innervation of the swim bladder in the zebrafish (Danio rerio). J. Comp. Neurol. 495: 587–606.

Frescura, F.A.M. and Hiley, B.J. 1984. Algebras, Quantum Theory and Pre-Space. Revista Brasileiara de Fisica Special Volume, 49–86.

Gaunt, S.J. 2015. The significance of Hox gene collinearity. Int. J. Dev. Biol. 59: 159–70.

Hawking, S. 2011. A Brief History of Time. Bantam, New York.

Hopwood, N. 2022. 'Not birth, marriage or death, but gastrulation': the life of a quotation in biology. Br. J. Hist. Sci. 55: 1–26.

Husserl, E. 1939. The origin of geometry. Rev. Int. Philos. 1: 157–179.

Kauffman, L. 1994. Knots and Physics. World Scientific Publishing Company, Singapore.

King, S.E. and Skinner, M.K. 2020. Epigenetic transgenerational inheritance of obesity susceptibility. Trends Endocrinol. Metab. 31: 478–494.

Lambertz, M., Grommes, K., Kohlsdorf, T. and Perry, S.F. 2015. Lungs of the first amniotes: why simple if they can be complex? Biol. Lett. 11: 20140848.

Mandl, P., Hayer, S., Karonitsch, T., Scholze, P., Győri, D., Sykoutri, D., Blüml, S., Mócsai, A., Poór, G., Huck, S., Smolen, J.S. and Redlich, K. 2016. Nicotinic acetylcholine receptors modulate osteoclastogenesis. Arthritis Res. Ther. 18: 63.

McGilchrist, L. 2019. The Master and His Emissary. Yale University Press, New Haven.

McMillen, M.M. 1979. Differential mortality by sex in fetal and neonatal deaths. Science 204: 89–91.

Popper, K. 1992. The Logic of Scientific Discovery. Routledge, London.

Purevdorj-Gage, B., Sheehan, K.B. and Hyman, L.E. 2006. Effects of low-shear modeled microgravity on cell function, gene expression, and phenotype in Saccharomyces cerevisiae. Appl. Environ. Microbiol. 72: 4569–4575.

Romer, A.S. 1949. The Vertebrate Story. Chicago, IL: University of Chicago Press, Chicago.

Rowlands, P. 2014. The Foundations of Physical Law. World Scientific Publishing Company, Singapore.

Sacco, R.G. 2019. The Fibonacci Life-Chart Method (FLCM) as a foundation for Carl Jung's theory of synchronicity. J. Analyt. Psychol. 61: 203–222.

Sagan, L. 1967. On the origin of mitosing cells. J. Theor. Biol. 14: 255–274.

Saintonge, J. and Côté, R. 1984. Fetal brain development in diabetic guinea pigs. Pediatr. Res. 650–653.

Sakurai, R., Cerny, L.M., Torday, J.S. and Rehan, V.K. 2011. Mechanism for nicotine-induced up-regulation of Wnt signaling in human alveolar interstitial fibroblasts. Exp. Lung Res. 37: 144–154.

Torday, J.S., Torday, D.P., Gutnick, J., Qin, J. and Rehan, V. 2001. Biologic role of fetal lung fibroblast triglycerides as antioxidants. Pediatr. Res. 49: 843–849.

Torday, J.S., Sun, H., Wang, L., Torres, E., Sunday, M.E. and Rubin, L.P. 2002. Leptin mediates the parathyroid hormone-related protein paracrine stimulation of fetal lung maturation. Am. J. Physiol. Lung Cell Mol. Physiol. 282: L405–L410.

Torday, J.S. 2003. Parathyroid hormone-related protein is a gravisensor in lung and bone cell biology. Adv. Space Res. 32: 1569–1576.

Torday, J.S. and Rehan, V.K. 2009. Exploiting cellular-developmental evolution as the scientific basis for preventive medicine. Med. Hypotheses 72: 596–602.

Torday, J.S. and Rehan, V.K. 2012. Evolutionary Biology, Cell-Cell Communication and Complex Disease. Wiley, Hoboken.

Torday, J.S. and Rehan, V.K. 2017. Evolution, The Logic of Biology. Wiley, Hoboken.

Torday, J.S. and Miller, W.B., Jr. 2018. Terminal addition in a cellular world. Prog. Biophys. Mol. Biol. 135: 1–10.

Torday, J.S. 2019. The Singularity of nature. Prog. Biophys. Mol. Biol. 142: 23–31.

Torday, J.S. 2020. Consciousness, Redux. Med. Hypotheses 140: 109674.

Torday, J.S. 2021a. Cellular evolution of language. Prog. Biophys. Mol. Biol. 167: 140–146.

Torday, J.S. 2021b. Life is a mobius strip. Prog. Biophys. Mol. Biol. 167: 41–45.

Walton, D.L., Norem, C.T., Schoen, E.J., Ray, G.T. and Colby, C.J. 1999. Second-trimester serum chorionic gonadotropin concentrations and complications and outcome of pregnancy. N. Engl. J. Med. 341: 2033–2038.

Wurm, G. and Teiser, J. 2021. Understanding planet formation using microgravity experiments. Nat. Rev. Phys. 3: 405–421.

Chapter 5
Cellular Evolution of Language as Quantum Energy

Introduction

We are at a critical inflection point in human history, with dwindling resources and huge pressures for food, water and medicine brought about by Climate Change in combination with Globalization of economies. We need an effective, testable and refutable narrative for human evolution so that we may rationally and predictably affect control of our destiny.

Cell-cell communication as the basis for evolution

A cellular-molecular approach based on cell-cell communication offers such a 'path through the forest', deconvoluting the evolution of vertebrate physiology, acting as a bauplan for understanding the human condition.

In that spirit, the physiologic evolution of language has been reconstructed, a difficult subject that has been bandied about formally ever since the publication of Darwin's "Descent of Man" in 1871. But literally all other attempts to understand the evolution of language have been descriptive and synchronic (same space-time), whereas the only way to understand human physiologic traits is to cross-cut and transcend space-time diachronically (across space-time) in order to factor out the human, subjective 'signature' that clouds our perspective, which David Bohm characterized as the Explicate Order in his book "Wholeness and the Implicate Order" (1980). Moreover, the holism that language is contingent on is the product of the local/non-local principle behind the Quantum Entanglement-Symbiogenesis interface.

By example, it has previously been shown that the evolution of the lung can be traced back to the unicellular state based on the functional genomic interrelationship between lipids, gravity and oxygen, mediated by cell-cell communications constrained by homeostasis, culminating in "The First Principles of Physiology" (Torday and Rehan, 2009). When that perspective is reversed, it predicts many physiologic traits that up until now have been seen as descriptive dogma rather than causally predictive explanations.

Using the same approach, language may also be deconvoluted based on reverse-engineering since it too is a physiologic trait. It is true that there are large gaps in our ability to trace the step-wise evolution of language, but that is largely because we have been thinking about evolution from its ends, not its means, its consequences, not its etiology, so scientists have not been formulating experiments along such lines up until now. But that can be corrected by formulating hypotheses based on cell-cell communication mechanisms for both ontogeny and phylogeny, as has previously been demonstrated.

The demonstration that the hormone leptin has the same developmental effect on the frog lung as it does on the mammalian lung (Torday et al., 2009) is experimental evidence for evolution as a continuum involving epigenetic change rather than random genetic mutations, the latter randomness eliminating the possibility for seeing the process of lung evolution as such a step-wise systematic approach based on homeostatic control, by definition.

It is of value here to point out that the debate over whether evolution is gradual (Darwin, 1859) or 'punctuated' (Eldredge and Gould, 1972) remains an open question. Yet by focusing on the cellular-molecular components of physiologic evolution that the process can be seen as both gradual and punctuated, depending upon how you perceive it. This is analogous with Bohr's Complementarity explanation for whether light is a particle or a wave. And in fact, the lowest level for the origin of physiologic evolution for gas-exchange, namely water-electrolyte balance in reference to language remains cryptic.....until it is seen in the context of the water-land transition, at which point the adaptation to land causes the overt phenotypic expression of these traits for the first time. Most of the terrestrially-adaptive traits—air breathing, micturition, skin barrier function, the brain—are accounted for by the duplication of the Parathyroid Hormone-related Protein Receptor; the terrestrially-adaptive evolution of language is linked to the duplication of the βAdrenergic Receptor during the water-land transition, which facilitated lung evolution [see Figure 1], the latter mediating bipedalism, freeing the forelimbs for toolmaking and oral language, the control of both being localized to the Area of Broca. Written language emerged as a result of the synergistic interactions between toolmaking and language.

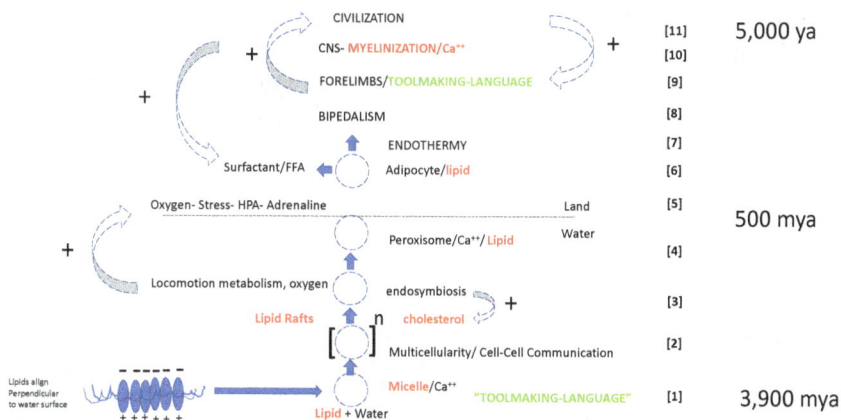

Figure 1. On the evolution of language. Time-Line in "years ago" (= ya) on the left-hand side. The formation of micelles [1] led to multicellular organisms (represented by the 'n' as an exponent) [2], the introduction of cholesterol into the cell membrane [3], the evolution of the peroxisome [4], the water-land transition and duplication of the βAdrenergic Receptor [5], the evolution of endothermy [6–7], leading to bipedalism [8], freeing the forelimbs [9], selection pressure for myelinization of neurons to facilitate calcium flux [10], culminating in the evolution of civilization [11].

The above step-wise integration of physiology has been hypothesized to be the basis for consciousness, particularly because it would explain David Chalmers' 'hard problem', and Krishnamurti's statement that 'the observer is the observed'. And these functionally-integrated interrelationships would also provide a rationale for the capacity of language to effectively express ourselves not only in daily parlance, but in literature, poetry and the lyrics for musical compositions alike, literally expressing our most visceral feelings and thoughts.

Bottom-up, top-down, middle-out

There are three modalities for evolution—Bottom-up, Top-Down, or Middle-Out. The first two are totally descriptive, lacking any mechanistic explanation. Darwinian evolution based on Natural Selection is a Top-Down process, but there is no mechanistically causal explanation, it's merely a metaphor that is untestable. Top-Down selection has also been referred to as 'downward causation', but that too is teleologic. Bottom-up selection is similarly descriptive and without a mechanistic basis other than Horowitz's 'one gene one enzyme' hypothesis (Horowitz, 1945), for which there is no empiric evidence. In contrast, the Middle-Out process is highly mechanistic, based on cell-cell communications mediated by soluble growth factors and their cognate receptors signaling for homeostasis, beginning with embryologic development, culminating in physiologic control; in this light, injury-repair is part of this continuum. This cellular-molecular approach can be exploited to understand the process of evolution based on falsifiable experimental evidence. And importantly, it is predictive, offering ways of solving such complex problems facing humanity as Climate Change and the practice of Preventive Medicine.

On language evolution as serial "middle-out" pre-adaptations

The term "middle-out" in the title of this section refers to the mechanism of cell-cell communication for embryologic structure and function, resulting in homeostatic physiologic control as the fundament for language [see Figure 1]. By contrast, all of the existing theories of language evolution refer to steps 9–11 in the Figure, seen 'horizontally', synchronically, while the current hypothesis refers to step-by-step interactions between cells and their environments epigenetically, from step 1–11 diachronically. The synchronic approach only offers associations and correlations, whereas the diachronic approach allows for causal, falsifiable hypotheses.

Referring to the accompanying Schematic, enumerated step-wise on the right-hand side: [1] Lipids are amphiphiles, meaning that they have either a positively or negatively charged "pole". When immersed in water, lipid molecules self-assemble perpendicular to the surface of the water under the influence of gravity, with their negatively charged poles pointed downward into the water (hydrophilic), and their positively-charged ends pointing upward into the air. In this configuration, the lipids will pack together, reducing the surface tension of the water surrounding their negatively-charged poles, spontaneously causing the quantum shift to the formation of spheres with semipermeable membranes due to the force of gravity. Such spheres are referred to as micelles, in effect forming the Explicate Order from the Implicate Order; [2] Subsequently, competition between eukaryotic and prokaryotic unicellular organisms, the latter forming pseudo-multicellular states such as biofilm and quorum sensing, causing true multicellular organisms to form by soluble growth factors signaling for structure and function, subsuming the initiation of life referred to in step [1] based on the First Principles of Physiology—negentropy, chemiosmosis and homeostasis; subsequently, cholesterol synthesis [3] was caused by the amount of oxygen in the atmosphere, one molecule of cholesterol requiring 11 atoms of oxygen. But the impetus for the synthesis of cholesterol in the face of larger amounts of oxygen based on serial pre-adaptations was derived from cellular cooperativity [1], and the exploitation of lipids to form the first cell [2]; later in vertebrate cellular evolution, calcium levels in the primordial ocean were gradually rising due to carbonic acid formed from carbon dioxide produced by plants leaching minerals from the bedrock. To cope with the existential threat of free calcium being toxic to lipids, cells stored calcium in their endoplasmic reticulum. But the calcium stores leaked out into the cytoplasm of the cell under oxidative stress, compensated for by the evolution peroxisomes [4], which utilize lipids to 'buffer' the toxic effect of excess calcium in the cytoplasm, thus recapitulating steps [1–3] in cellular evolution; subsequently [5], the hypoxic stress caused by the stepwise evolution of the lung enlisted the previous 4 biologic traits, mediated by the increased production of adrenaline, causing increased secretion of lung surfactant by the alveoli (Lawson et al., 1978), facilitating increased oxygenation by acutely allowing greater expansion of the alveoli. An inadvertent 'side-effect' [6] of hypoxic stress-induced adrenaline production was the breakdown of adipocytes, releasing free fatty acid into circulation, increasing metabolism, producing more metabolic heat as the earliest form of endothermy [7]; again, this trait was contingent on the previous 6 listed enumerated traits; endothermy [8] facilitated bipedalism since standing on two legs requires more efficient metabolism for energy production than quadrupedalism does. Berwick and Chomsky's "Why Only Us" hypothesizes a small growth factor signaling change as the missing link from chimpanzees to humans; the roles of FoxP2 in development of lung, gut and bone as pleiotropies implicates Foxp2 in bone adaptation for bipedalism unique to humans, i.e., selection pressure for Foxp2 links bipedalism with language formation. Subsequently, bipedalism gave rise to freeing of the forelimbs for toolmaking [9], placing positive selection pressure on the myelinization of neurons in the central nervous system [10], as a consequence of steps [1–9] (parenthetically, the homology with skin barrier formation mediated by Neuregulin forms the basis for these linkages); and because the hands were pre-occupied with toolmaking, positive selection pressure for bipedalism also gave rise to language [11]. Evidence for the co-evolution of motor skills and language lies in the fact that the neuronal basis for both of these functions resides in the cerebral Area of Broca—all Great Apes possess an Area of Broca, but only humans are bipedal, generating the positive selection pressure for both toolmaking and language. Further synergistically positive selection pressure on toolmaking and language led to further adaptations underpinned by steps [1–10], giving rise in the aggregate to civilization and culture.

Thus, as a derivative of toolmaking, language forms the basis for all human endeavor—literature, music, art, social sciences, geology, economics, sciences, psychology, etc., as transfers of energy. As such, the mechanism for the evolution of language can form the prototype for deconvoluting the other disciplines, forming the basis for a Periodic Table of Education. This would constitute a paradigm shift in the way we think about communicating knowledge, not as material things, but as a faithful transfer of energy. In his book "The Structure of Scientific Revolutions", Kuhn said that the hallmark of a paradigm shift is a change in the language, providing a rationale for focusing on language. And since language is hypothesized to be derived from toolmaking, it helps to deconstruct other disciplines such as literature, for example, offering ways that lead to deeper understanding of the human condition through narrative; perhaps even closer to the core concept of toolmaking than prose is poetry, which is more inferential than descriptive, alluding to the Implicate versus the Explicate Order. For example Emily Dickinson's characterization of "hope" as "the thing with feathers", merges verse with biology as a unity. Or Robert Frost, saying that "Life is that which can mix oil and water", or as his poetry expresses the enigma in "The Secret Sits":

We dance round in a ring and suppose,
But the Secret sits in the middle and knows.

Or art as a way of expressing abstract ideas, like Henry Moore's sculptures with large holes through them [see Figure 2], asking us to deliberate whether it is the material portion of the sculpture or the negative space that is of primary significance, or both? Or economics as the materialization of human physiology, begging the question whether it is 'supply and demand' or 'Gibbs free energy' that is more faithful to the discipline?

Key to understanding the seemingly directed evolution of language, as portrayed above, is that the formation of the first cell by lipids allowed for the formation of the First Principles of Physiology—negative entropy, chemiosmosis and homeostasis. All of the subsequently evolved traits reference the maintenance and perpetuation of those principles.

Figure 2. Henry Moore sculpture. Note the hole, or negative space.

Self-referential self-organizing self-authorship

Vertical integration of cellular evolution as the process by which language has evolved leads to the question as to where it originated from? The answer lies in the first step in the progression, the immersion of lipids in water, aligning at the water-air interface, their negatively charged ends pointed downward into the water because they are hydrophilic, their positively-charged ends pointing skyward. And the tighter the lipids pack together the more effectively they reduce the surface tension of the water, enabling the formation of micelles or primitive protocells. In so doing they form the Explicate Order that Bohm describes in his book Wholeness and the Implicate Order, that fictionalized version of the totality we have exploited in order to evolve our state of being.

Importantly, micelles float at the surface of the ocean because of their low density, resulting in being warmed by the Sun by day, liquifying and deforming at night. In the absence of the Sun, the micelles recapitulate their original form since they exhibit hysteresis, or molecular memory. That recursive warming and cooling tended to trap more and more calcium ions within the micelles, presenting an existential problem since calcium is toxic for lipids, causing them to denature. However, a subset of such micelles hypothetically formed calcium channels that mediated the uptake and release of calcium ions, facilitated by the proportional uptake and release of water molecules due to osmosis. That sequential uptake and release of calcium and water would have generated the pulsatility characteristic of both unicellular organisms and multicellular organisms alike. But perhaps more importantly, the dynamic hysteretic interrelationship between lipids and calcium was the origin of the molecular memory critical for evolution, the organism needing to remember its previous existential encounters by which it coped in order to combat successive life or death encounters.

A classic example of such molecular memory is cholesterol synthesis, which Konrad Bloch, who discovered the biosynthetic pathway for cholesterol, referred to as a 'molecular fossil'. He rationalized that since it takes 11 atoms of oxygen to produce one molecule of cholesterol, there had to have been ample oxygen in the atmosphere to do so. However, that was illogical reasoning after the fact. Based on the serial pre-adaptation approach, the advent of cholesterol referenced the exploitation of lipid to form life in the first place. The insertion of cholesterol into the cell membrane advanced eukaryotic evolution, thinning the phospholipid bilayer, increasing oxygen uptake, metabolism, and locomotion, the latter being due to increased cytoplasmic streaming. The presence of cholesterol in the cell membrane also formed the basis for the lipid rafts from which cell-surface receptors generated intercellular signaling to form multicellular organisms. The inception of these properties predicted many future vertebrate traits.

It was during the water-land transition some 500 million years ago that such lipid-related traits again proved existential. The β-Adrenergic Receptor duplicated, amplifying that signaling pathway throughout the organism. Because the lung was evolving from the swim bladder in the effort of fish to adapt to land, there were episodes of physiologic stress due to the inadequacy of the nascent lung to provide enough oxygen, hypoxia being the most potent stressor known in vertebrate physiology. In response, the hypothalamic-pituitary-adrenal axis was activated, stimulating secretion of Adrenocorticotrophic Hormone (ACTH), which stimulated adrenocortical production of cortisol, which in turn stimulated the production of adrenaline by the adrenal medulla. Adrenaline stimulated the secretion of surfactant by the alveoli, making them more distensible, acutely increasing the amount of oxygen in circulation, relieving the constraint on lung evolution; over the long-haul the distension of the alveoli caused increased secretion of Parathyroid Hormone-related Protein, which in turn generates more alveoli. As a side-effect, adrenaline caused the rupture of fat cells, releasing free fatty acid into the circulation, increasing metabolism and body heat. That was the adaptive origin of warm-bloodedness, or endothermy.

The advent of endothermy allowed for bipedalism, or walking on two legs, since bipedalism requires much more energy than walking on four legs. And given the pre-adaptive referencing of earlier traits for emergence, the upright posture may have been in reference to the perpendicular alignment of lipid molecules at the inception of life, particularly given the significance of bipedalism in freeing the forelimbs for toolmaking. As an intermediate step in the progression for bipedalism, eukaryotes evolved deuterostomy, or development from the anus to the mouth, causing selection pressure for Terminal Addition, underpinned by vagal evolution.

The manufacture of tools may have been the origin of language from a practical standpoint since the hands were preoccupied with toolmaking, otherwise hampering manual gesticulations for the purposes of communication. Conceptually, the forming of an arrowhead from a piece of flint is not very different from conceiving of a thought and executing it by adding a subject and object to a verb. Positive selection pressure for the combined action of hands for toolmaking and language increased myelinization of neurons in the brain, enhancing mental capacity. The synergistic interaction of toolmaking, language and central nervous system gave rise to written language, fostering the creation of Civilization.

Holism

Ontologically, the Big Bang provided a point source for the origin of the Universe, first detected as the microwave background that echoed that explosive event. In a recent publication, a homologous point source was identified as the unicellular origin of life. This is not an analogy or metaphor—the continued existence and evolution of eukaryotes was ensured by the biosynthesis and insertion of cholesterol into the cell membrane, facilitating vertebrate metabolism, locomotion and respiration, the three foundational traits of vertebrate evolution. Consequently, the cell membrane became more fluid, enabling endocytosis and exocytosis, allowing for freer access of environmental factors to the cytoplasm for substances that would otherwise have killed eukaryotes off long ago—heavy metals, ions, gases—instead compartmentalizing them with endomembranes, rendering them useful for physiologic functions.

Thinking from the unicellular state forward, biology interacting with the ever-changing environment, over and over again, the innate continuum of physics and biology becomes self-evident. It is a means of understanding Gaia, James Lovelock's "Earth Mother" concept as the cell continuously internalizes the environment to generate niche constructions.

Organisms meet environmental challenges by reallocating genetic traits previously used in their evolution. For example, the emergence of cholesterol in the cell membrane was an exaptation characteristic of the overall process of vertebrate evolution. In service to homeostasis, lipids were utilized as anti-oxidants, lipid rafts for ligand-receptor cell-cell signaling, and as substrate for the steroid hormones of the endocrine system. This is referred to as the Endosymbiosis Theory, which applies to all eukaryotes, from protozoa to Man, and every organism in between. By internalizing the environment, organisms have adapted to it, evolving internal organs over the course of vertebrate evolution. In parallel, sensing mechanisms, ranging from the unicellular cell membrane to evolved organs of sensing have been vertically integrated, culminating in the nervous systems of more complex organisms. In the aggregate, this iterative process is consciousness, or mind, as the way in which we intuit our surroundings, which is radically different from conventional ways of thinking about consciousness either as being in our heads or extending into the environment.

Cell-cell communication as 'parts-to-wholes' language evolution

Given that the Laws of Physics determine all of the Natural Laws begs the question how physics determines biology. Physicists have tried to resolve this question and failed, concluding that biology is just too complicated. But that may be because they have looked at the problem from its results instead of its progression; in contrast to that experience, there are homologies (common origins) between physics and biology emanating from their common origins in the Singularity which gave rise to the Big Bang. For example, Smolin has determined that stellar evolution and Black

Holes can be explained based on Darwinian evolutionary mechanisms. In the aftermath of the Big Bang, it is feasible based on Newton's Third Law of Motion, that an 'equal and opposite reaction' formed as homeostatic self-referential self-organization. Chemistry and biology may have emerged on that basis, though it can be argued that we cannot know the origins of the totality since we were not present when it occurred. However, we can surmise how and why it happened based on scientifically testable and refutable hypotheses. It has been argued elsewhere that since lipids were critically important in the evolution of eukaryotes, they putatively played a vital role in the origins and initial conditions of life, evolution being pre-adaptive, or exaptive, or co-optional. Lipids immersed in water may have formed the basis for life since they accompanied the frozen snowball-like asteroids that formed the Earth's ocean, and physicochemically spontaneously form micelles, or primitive cells, when immersed in water, exhibiting hysteresis as molecular 'memory', able to recall their shape and size despite being deformed and reformed iteratively, which is necessary for evolution—the lipid membranes that demarcated the internal and external environments generated life as an ambiguity. Under these conditions, several homologs of Quantum Mechanics apply—namely the Pauli Exclusion Principle, non-localization, the Heisenberg Uncertainty Principle and Coherence. In light of the First Principles of Physiology, For example, with respect to the Pauli Exclusion Principle, the first three quantum variables are deterministic, whereas the fourth is probabilistic. Thus, life exists between the boundaries of Determinism and Free Will. So, biology complies with physics as the co-existence of both Determinism and Free Will; biology, in turn, mimics the Quantum Mechanical principles through homologies with the First Principles of Physiology. Similarly, both biology in the form of pleiotropy, and the distribution of the elements in the totality demonstrate that Quantum Mechanics and biology are non-localized. Pleiotropically, under physiologic stress, the same gene may be expressed in a variety of tissues, but always with reference to the originating unicellular First Principles of Physiology, distributed throughout the organism in a non-localized manner. This distributive property is homologous (having the same origins) with the non-localization principle of Quantum Mechanics. In his book What is Life? Schrodinger stated that the internal free energy or entropy of the cell is negative, in contrast to the positive entropy outside of the cell, so life began as an ambiguity; that ambiguity, or uncertainty, is the on-going driver for the subsequent evolution of the organism over time, both developmentally and phylogenetically. For example, Niels Bohr explained the seemingly paradoxical duality of light as both particle and wave being an artifact due to the differences in the way that it is observed, which is homologous with the ambiguity of the cell, coping with such paradoxes, as described above; perhaps it is because physics, like life itself exists in a state of limbo regarding its surroundings that allows it to contend with and contemplate it, even at the Quantum Mechanical level — *like dissolves like*.

In order to become a 'hard' science, biology needs the prediction of Quantum Mechanics for evolution. What follows is a way of understanding biology mechanistically, allowing it to interface with Quantum Mechanics functionally, revealing the common source of life as the Singularity/Big Bang. By internalizing the environment through endosymbiosis, the unicell, was post-dicted to be the first Niche Construction, physiology made functional by compartmentalization; the ensuing communication between cells in service to metabolic cooperativity was realized through Niche Construction, in combination with the inheritance of epigenetic marks, giving rise to larger and larger organismal communities. Widespread systematic dissemination of these collective properties gave rise to Gaia, the concept of the organic unity of the Earth. Fluctuations in the levels of oxygen and carbon dioxide in the Earth's atmosphere have caused physiologic stresses on life forms over the course of the Earth's history. Based on that precept, the evolution of the mammalian lung was traced from its unicellular origin forward by aligning the phenotypic changes in the gas exchanger with the ups and downs of oxygen in the atmosphere; using that approach revealed that evolution could literally be viewed homologously with Quantum Mechanics (QM). Furthermore, by reducing the origin of life to its elements, expressed as the First Principles of Physiology enables seeing the one-to-one relationships between QM and physiology, much like Mendeleev's Periodic Table, reducing the hierarchy of the Elements to the number of protons in their nuclei. Consequently, QM can be applied directly to the process of evolution as a homology with the Pauli Exclusion Principle, non-localization, or Heisenberg Uncertainty Principle. For example, Pauli Exclusion Principle is the QM principle that no two electrons can have the same spin. Electron spin is determined by its four quanta of energy, the first quantum determining the quantum of the second and third; the last quantum number is time-dependent, and is therefore probabilistic. Therefore, each electron exists between the boundaries of determinism and probability. The cell similarly resides between negentropy and chemiosmosis, which are deterministic, and the Free Will of homeostasis, which is probabilistic. Another example of the homologies between biology and QM is the non-local distribution of physical properties throughout the totality of the Cosmos, which produces pleiotropic traits by acquiring the same gene over the course of its history under different environmental conditions.

Fractals, 'parts of the whole'?

The cell may be seen as a fractal of physiology, and beyond that to the Cosmos. Seen in that light, language, as a middle-out expression of cell-cell communication 'verbalizes' the vertical integration from the unicell to consciousness. This way of thinking helps to explain such phenomena as Near Death Experiences, Out of Body Experiences, and Maslow Peak Experiences as natural consequences of the synchronous activation of cell-cell communications, culminating in unified 'fields' of calcium waves. As such, these experiences allow for the convergence of our individual consciousness with that of the Cosmos as the totality of Natural Laws.

Top-down, bottom-up or middle-out control of physiology

The question of whether physiology is top-down or bottom-up has been debated for decades. At issue is what the basis for such control systems is. The middle-out approach invoked herein for understanding the evolution of language begins with the premise that structure and function are determined by cell-cell communication, mediated by growth factor/receptor signaling that begins at the inception of life in the zygote and persists until death. Its origins can be traced back to the unicellular state, and beyond that, to the Singularity. And because of its validity, it provides a number of explanations for dogma—the nature of the cell, the function of homeostasis, pleiotropy, the life cycle, heterochrony—all of which are untenable based on descriptive biology.

Discussion

The field of linguistics has a long history, dating back formally to Darwin's "Descent of Man", in which he speculates that "the survival of certain favoured words in the struggle for existence is natural selection". From there on out, theories of language have strictly been set in the frame of its overt, synchronic characteristics, reasoning after the fact. In contrast to what has come before, the present hypothesis that language evolved from the process of cell-cell communication, which determines embryogenesis, physiologic homeostasis and injury-repair, cuts across space-time diachronically, eliminating the encumbrances of anthropocentric thinking. Instead, by focusing on language as a natural consequence of a hierarchically derived, holistic integration of the cell and its environment offers a novel, *a priori* way of understanding the true origin and process of linguistics that is testable and refutable rather than being subjectively inductive and speculative.

A systematic understanding of linguistics is critically important as we move into the Anthropocene, or a man-made world, accepting the reality of Artificial Intelligence without mechanistically predictive knowledge of our evolution. We are tampering with aspects of human traits that we do not understand ontologically and epistemologically, potentially exposing us to the foibles of our ignorance through self-deception to the point of being blinded to the error of our ways. The classic example is the Anthropic Principle, which takes the view that we have been put here and have had the great good fortune to end up in a place that accommodates our physiology. In reality, we are not 'in' this place, we are 'of' this place, literally. Carl Sagan used to sign off on his Cosmos TV show saying that "The cosmos is within us. We are made of star-stuff. We are a way for the universe to know itself". He intuited this, lacking experimental evidence, but was probably being influenced by his wife at the time, Lynn Margulis Sagan, who championed the Endosymbiosis Theory, predicated on the idea that we have evolved by internalizing things in our environment that would otherwise have destroyed us. So, in the aggregate she was providing the scientific evidence for Sagan's trope. It took the discovery of soluble growth factors and their receptors as the basis for embryogenesis and physiology to understand the underlying mechanism.

Nowhere is the value of the above perspective more evident than in the deconvolution and recapitulation of the evolution of language as expressed herein. Language emerges from the bowels of the Cosmos, forming a logic observable in the Periodic Table of Elements, the Elements distributed by the stars as they generated light from matter, beginning with the lightest, hydrogen, and ending with the heaviest and most proton-rich elements. The Cosmos is 13.8 billion years old, and it took 9 billion years until the Earth was formed. The first cell appeared once the ocean formed from snowball-like asteroids pelting the planet, the polycyclic hydrocarbons that were also present spontaneously forming micelles, or prototypical 'cells'. The resulting delineation of 'inside' and 'outside' was the origin of language as a 'dialogue' between life and non-life, which culminated in oral and written language, the latter being the key to the rise of civilization. It is imperative that we understand this intimate relationship between our physiology and our environment through language as the 'canary in the coalmine'. We must free ourselves from the deception that Robert Trivers has determined is our un-doing in his book "The Folly of Fools". To do otherwise puts us at risk of extinction.

References cited

Bohm, D. 1980. Wholeness and the Implicate Order. Routledge and Kegan Paul, New York.

Darwin, C. 1859. On the Origin of Species. John Murray, London.

Eldredge, N. and Gould, S.J. 1972. Punctuated Equilibria: An Alternative to Phyletic Gradualism Models in Paleobiology. Cooper & Co, San Francisco.

Horowitz, N.H. 1945. On the evolution of biochemical syntheses. Proc. Natl. Acad. Sci. USA 31: 153–157.

Lawson, E.E., Brown, E.R., Torday, J.S., Madansky, D.L. and Taeusch, H.W. Jr. 1978. The effect of epinephrine on tracheal fluid flow and surfactant efflux in fetal sheep. Am. Rev. Respir. Dis. 118: 1023–1026.

Torday, J.S., Ihida-Stansbury, K. and Rehan V.K. 2009. Leptin stimulates Xenopus lung development: evolution in a dish. Evol. Dev. 11: 219–224.

Torday, J.S. and Rehan V.K. 2009. Lung evolution as a cipher for physiology. Physiol. Genomics 38: 1–6.

Chapter 6
Fibonacci Numbers, Self-Reference, Self-Organization and Autopoiesis*

Introduction

Maturana and Varela are renowned for describing life as 'Self-Referential and Self-Organizing', or Autopoiesis (1972), but without a mechanistic understanding of what that is, there is no possibility for empiric testing, let alone a deeper understanding of evolutionary causation.

Enter the discovery that development/embryology is mediated by soluble growth factors signaling through their cognate receptors. That was a game-changer, leading to a comprehensive understanding for the well-known transition from the fertilized egg, or zygote, to the subsequent series of cell divisions ultimately giving rise to the off-spring. But there was more to that than met the eye, as was pointed out by Lewis Wolpert, who stated that "gastrulation was the most important thing you would ever do in your life", which was provocative, but has remained enigmatic since Wolpert did not explain the basis for his statement. Gastrulation is the stage in embryologic development when the mesoderm is introduced as the third germ layer. It is of vital importance in providing plasticity to the developing conceptus. Moreover, it is under epigenetic control, providing a way of understanding 'how and why' it is of such great importance in evolution as the capacity for adaptive change. And it should be borne in

---

* Written in collaboration with Robert Sacco, PhD.

mind that the cell is a fractal, as described in the introductory chapter, acting to forge the link between the inanimate and animate.

The dynamism of the mesoderm is best seen during the transition of vertebrates from water to land. That process occurred when Romer's 'Greenhouse Effect' caused the partial drying up of the oceans, exposing land masses around the globe. Equally important in the evolution of land vertebrates was the depletion of oxygen from the water, given that the partial pressure of a gas dissolved in water is contingent on temperature. The combined effects of dry land and less oxygen in water drove a specific subset of boney fish onto land, revealing the evolutionary strategy of using pre-existing traits. Physostomous boney fish are those with a pneumatic duct connecting their swim bladder to their esophagus. It is known that there were at least five independent attempts by Physostomous fish to inhabit land, suggesting some sort of a step-wise 'trial and error' scenario. During that transition, three gene duplications occurred — the Parathyroid Hormone-related Protein Receptor (PTHrPR), the Glucocorticoid Receptor (GR) and the ßAdrenergic Receptor (ßAR). Please note that all three duplications involved hormone receptors, which are by definition amplification mechanisms for cell-cell signaling.

The further amplification of such receptor-mediated mechanisms increased cell-cell signaling exponentially, but to what end? Each of those receptors was existential for the survival of vertebrates on land. The duplication of the GR was essential initially in order to offset the Mineralocorticoid (MR) effect on blood pressure since the physiologic stress of the transition from water to land increased blood pressure due to the increase in the effect of gravity, and the MR further increased blood pressure, which became life-threatening. The emergence of the GR from the MR offset the latter's effect on blood pressure. The addition of three amino acids to the binding site of the MR gave rise to the GR, likely by 'trial and error' since there is no innate effect of such amino acids other than in the context of the role of the MR in blood pressure control.

Furthermore, there was synergy in the convergence of the advent of the GR and the expression of ßAR's, particularly in the microvasculature of the evolving lung, giving rise to the independent control of pulmonary blood pressure. The lung evolved from the swim bladder of Physostomous boney fish, the pneumatic duct connecting the esophagus to the bladder being the homolog of the trachea. The causal nature of this event is underscored by the duplication of the nicotinic Acetylcholine Receptor (nAChR) in boney fish, acting to augment calcium flux in the smooth muscle sheath around the pneumatic duct/trachea in tandem with the enhancement of bone remodeling in the fins, allowing them to facilitate the movement out of water onto land. That process was aided by the duplication of the PTHrPR, which mediates the calcification of bone, providing the structural evolution of the fins in support of Tiktaalik, the quadrupedal transition vertebrate.

In combination with the role of PTHrP in locomotion on land, PTHrP also had land-adaptive effects on the swim bladder, glomerulus and skin, all of which are inhibited experimentally by deleting the PTHrP gene. In the swim bladder, PTHrP is necessary for the formation of alveoli; in the glomerulus it is necessary for movement of fluids and electrolytes into the descending tubules of the kidney; and in the skin, it is necessary for barrier formation.

The role of the nAChR in lung physiology came to light while studying the effect of cigarette smoke on childhood asthma. Nicotine is one of the components of cigarette smoke and its role in childhood asthma epidemiologically is well documented. Treatment of pregnant rats in mid-pregnancy caused asthma in the offspring for at least 3 more generations. The mechanism of action was enhancement of the nAChR in the smooth muscle of the trachea, increasing the flow of calcium through the smooth muscle, causing it to be more sensitive to cold air and particulates. This effect is in reference to the role of the nAChR in bone remodeling for the adaptation to land.

How/Why the nicotine epigenetic 'mark' is enhanced transgenerationally is somewhat obscure, but may hypothetically be due to its recognition as the Fibonacci sequence for the nAChR DNA code in the egg and sperm. The consequent enhancement of that code may be a sign that it should be methylated, resulting in increased expression of that gene in the offspring.

Nicotine enhances short term memory in the brain due to increased nAChR expression as well. The increased calcium flux in neurons enhances short-term memory. In combination with its effect on bone remodeling, duplications of the nAChR would have facilitated the well-recognized increase in cranial size along with the increase in brain volume.

Self-organization derives from the effect of gravity. Initially, gravity put pressure on lipid molecules floating in the ocean, causing them to orient vertically, their hydrophylic negative poles pointing downward, their hydrophobic positively charged poles pointing upwards. Such lipid molecules packed together, and when their negative charge reached a critical mass, it neutralized the Van der Waals force for surface tension, causing the lipids to spontaneously form micelles, or protocells. Such micelles acquired memory due to their hysteretic deforming and reforming due to the warming of the sun by day, and cooling by night. This self-organizing system was the source of life. Evidence for the validity of this characteristic has been demonstrated by placing cells in microgravity, resulting in their loss of phenotype or self as the manifestation of self-organization. Subsequent consequences of this trait such as deuterostomy, bipedalism and freeing of the forelimbs were all manifestations of such self-referencing self-organization.

Cell membrane as mobius strip

The above explanandum for the Self-Referential Self-Organization of life leads to post-dictions such as curiosity and imagination. Those features derive from the nature of the cell membrane as a partial barrier to particles flowing in an out, and the role of Symbiogenesis as the arbiter of what would be assimilated in our physiology. That emergent property was a consequence of Quantum Entanglement being 'energized' by gravity impinging on the protocell as the local effect, referencing the non-local gravity of the Cosmos. Moreover, because the cell could recall when it wasn't due to its referencing of the Cosmos, it acquired imagination, further selecting for all of the above characteristics holistically.

It is that holism that gave rise to human evolution. The non-local referencing of the Cosmos links Fibonacci sequences to the life form, as does Knot Math. Symbiogenesis causes the assimilation of life-threatening factors in the environment, but mathematics is not among them. Husserl teaches us that math in the guise of geometry exists external to us. Similarly, the cell is a homolog of the knot, the latter having to 'unknot' in order to prove its authenticity. In a sense the cell has to do the same, having 'knotted' itself to form structures and functions in order to monitor the environment for epigenetic marks. In order to reproduce though, the knots must unknot during meiosis in order to recapitulate the unicell as egg and sperm.

The evolution of the lung from the swim bladder referred to above was necessary for life on land. Terrestrial life placing increased stress on metabolic efficiency fueled by the lung, which intermittently caused hypoxia, the latter being the most potent stimulus for adrenalin. Adrenalin alleviated the limitation on the alveoli transiently by causing the increased secretion of lung surfactant, increasing the distensibility of the alveoli, increasing their surface area. As a side-effect, adrenalin caused secretion of free fatty acids from fat cells, providing highly efficient fuel for metabolism, increasing body temperature. The latter gave rise to endothermy, or the self-contained control of body temperature, which is much more efficient than being cold-blooded. Metabolic activity in warm-blooded organisms requires only one enzyme, whereas in cold-blooded organism it requires several steps in order to catalyze the reaction at different ambient temperatures efficiently. That led to the transition from cold-blooded to warm blooded organisms, giving rise to bipedalism, freeing the forelimbs for further evolution of the central nervous system in support of toolmaking and language. Imagination fostered the synergy between language and toolmaking, giving rise to civilizations that facilitated the evolution of humans who were highly dependent upon social systems in order to survive and thrive, given their superannuated adolescence.

Human evolution is characterized by an ever-enlarging skull and contents/brain. That is, until the head got too big to fit through the birth canal, leading to our forebears who evolved to birth prematurely, with only about 25% of total adult brain volume. It takes decades for the human brain to reach its full potential, necessitating an elaborate social support system—family, community, State, Nation State. Meanwhile, we tend to exhibit adolescent behaviors such as narcissism, risk taking, and being overly sexualized due to the lack of the usual constraints on such activity.

Cell membrane, memory and consciousness

In the aggregate, all of the above constitutes consciousness, beginning with the cell membrane acting as a selection mechanism for what enters and exits the cell. That process then provides the basis for physiology, mediated by cell-cell communication. Although the cell membrane provides the daily decision-making as binaries, the more challenging decisions are kicked upstairs to consciousness as cellular communication. Experimental evidence for that is provided by the Libet Experiment, showing a 300 millisecond gap between sensing something and reacting to it. In a more recent experiment, a similar observation was made. And under duress, it has been hypothesized that a 'wave collapse' occurs, causing convergence of the two levels of consciousness. And similarly, meditation can accomplish the same result without stress. In any case, the local nature of our memory can connect with the non-local of Cosmic Consciousness.

References cited

Maturana, H.R. and Varela, F.J. 1972. Autopoiesis and Cognition. Reidl, Boston.

Chapter 7

Jung's Synchronicity, Unicellular Consciousness, Quantum Mechanics and Epigenetic Inheritance

Introduction

One of us (J.T.) lived in Montreal, Canada during the 1960's and 70's, going to graduate school at McGill University. It is said that the weather there is '8 months of cold, followed by 4 months of bad sledding'. 'Round about February 'cabin fever' sets in', which is depressing due to the cold and grey weather. To cope, I would go to our kitchen pantry, unscrew the cap of the Thyme bottle and take a whiff. It smelled like the woods of the Berkshires, where I spent every summer growing up. My blues tended to dissipate, remembering those halcyon days. It is experiences like that make me wonder about consciousness/mind and their relationships to physiology and the Cosmos. But then the question to me as a scientist is 'how to deconvolute that?'. I will relate my reduction of consciousness in this way, and speculate as to how to exploit it to foster Larry Dossey's "One Mind" (2014, Hay House, Carlsbad).

Reductio ad Explanandum, or How emergences came to be

It was Darwin who initially said that evolved traits can come about from pre-conditions. That idea was exploited in combination with known mechanisms of cell-cell communication in order to deconstruct the evolution of the lung (Torday et al., 2007), followed thereafter by other complicated structures and functions, all of which biologically culminate in the unicell (Torday and Rehan, 2017). That raised the question as to how and why the cell evolved?

The evolution of the cell—a working model

Knowing that lipids and water were delivered to the primordial Earth about 100 million years after it cooled, it was hypothesized that the lipids formed micelles or protocells. In combination with Lynn Margulis Sagan's Symbiogenesis hypothesis (1967)—that we have evolved by assimilating factors in the environment that have posed existential threats—was consistent with both development and phylogeny. And since both processes can be reduced to cell-cell signaling mechanisms, it was feasible to provide scientific evidence to key steps in vertebrate evolution, both forward (Torday and Rehan, 2012) and in reverse (Guex et al., 2020).

Quantum Mechanics as the Ultimate Basis for Evolution

Having been able to explain several otherwise dogmatic concepts of biology using the reduction of emergences, emboldened speculation as to why the unicell gave rise to Symbiogenesis? Consider the process by which lipids form micelles, aligning at the surface of the primordial ocean, negatively-charged hydrophilic pole pointing downward, positively-charged hydrophobic pole pointing upward due to the force of gravity. Once a critical mass of such lipid molecules forms, it can neutralize the surface tension due to Van der Waals force, and the lipids make a quantum leap from individual molecules to a micelle, a lipid-enclosed sphere with a semi-permeable surface that dictates the passage of particles in and out of the cell. That condition subsequently evolves into the physiology of the cell through Symbiogenesis, providing both a mundane in-out kind of consciousness along with a continuum of physiologic consciousness. It is this two-tiered system of consciousness that determines our behavior.

By employing Einstein's Field Theory (1961) that when gravitational force is applied to a curved surface it produces energy. Such energy would have fostered and sustained the Quantum Entanglement of particles within the evolving cell locally to maintain and perpetuate homeostasis as one of the major tenets of Quantum Mechanics. In turn, that raised the question as to how and why Quantum Mechanics arose? And it occurred to me that hypothetically the cell membrane could be a functional Mobius Strip because the cell 'remembers' when it was lipid molecules floating on the primordial ocean that covered the nascent Earth. I would submit that is the basis for imagination as an extension of consciousness, enabling us hominins to conceive of Quantum Mechanics.

Evolution of imagination

I hypothesize that all organisms are conscious because they evolve by Symbiogenesis, but Man is unique in having imagination due to the way in which our consciousness has evolved due to endothermy enabling bipedalism, freeing the forelimbs for toolmaking in combination with language, oral and written as forms of toolmaking. That concatenation of traits put huge selection pressure on our central nervous system, giving rise to the brain as a specialized neuronal structure. Hence our capacity for imagination was enabled by bipedalism and endothermy.

Locomotion, imagination and epigenetic inheritance as the origin of Homo sapiens sapiens

When consciousness is seen as a continuum from lipid-based micelles with hysteretic memory, to endothermy/warm bloodedness, to bipedalism, to forelimb freedom, to toolmaking/language, to imagination, all of which reference the First Principles of Physiology, a rich understanding of what intelligence as contextual meaning emerges—in the context of the unicell, lipids as antioxidants were contextually adaptive; in the context of aquatic life, the swim bladder was adaptive; fast forward to Romer's 'Greenhouse Effect' and the morphing of the swim bladder into a lung was adaptive. It also accounts for the transition from 'fight or flight' in the wild to problem solving within a social system, for example, though that can breakdown maladaptively. It also accounts for 'qualia' as the aggregate of all of those characteristics of consciousness, whereas in the conventional way of understanding consciousness qualia are anecdotal and enigmatic. So David Chalmers' challenge to understand why we see 'red' when we injure ourselves becomes tractable as an atavism, referring to our 'memory' of bleeding as a new trait, for example.

We retain such memories in order to be able to nimbly use such traits effectively. They are exemplified descriptively as pleiotropy (Torday, 2018), and mechanistically as 'Phantom Limb', the retention of 'feeling' in a severed limb allowing for the up-stream components of said limb to remain functional in order for the organism to remain patent. That interpretation of an otherwise anecdotal view demonstrates the advantage of seeing evolution from its origin instead of after-the-fact.

Epigenetic inheritance as the means for sustaining life

The reason why it is critically important to sustain signals up-stream of a severed limb as described above for Phantom Limb is in order for the organism to remain phenotypically intact in order to detect abnormalities in the environment, given that the phenotype is the 'agent' for doing so (Torday and Miller, 2016). Loss of phenotypic characteristics would jeopardize the organism's purpose in detecting changes to an otherwise adapted environment, as recalled by the memory imbedded in the physiologic signaling for development and homeostasis.

Such imperfections in the environment are referred to as epigenetic 'marks'. In some unknown way, they are recognized as 'foreign' to the history of the organism. Marks are absorbed by the organism and processed within the germ cells. If the mark is determined to be of significance to the organisms history, it will be assimilated as a biochemical adduct-methylation, ubiquitination, etc.—resulting in a change in the transcription and translation of the DNA affected.

The mechanism of epigenetic inheritance is aided by Niche Construction, or the coordination of external and internal environments through Symbiogenic assimilation. In this way, the organism is highly attuned to changes in the environment, allowing it to detect and assimilate them in order to adapt effectively.

References cited

Einstein, A. 1961. Relativity. The Special and General Theory. Crown Publishing, New York.

Guex, J., Torday, J.S. and Miller, W.B. Jr. 2020. Morphogenesis, Environmental Stress and Reverse Evolution. Springer Nature, Switzerland.

Sagan, L. 1967. On the origin of mitosing cells. J. Theor. Biol. 14: 255–274.

Torday, J.S., Rehan, V.K., Hicks, J.W., Wang, T., Maina, J., Weibel, E.R., Hsia, C.C., Sommer, R.J. and Perry, S.F. 2007. Deconvoluting lung evolution: from phenotypes to gene regulatory networks. Integr. Comp. Biol. 47: 601–609.

Torday, J.S. and Rehan, V.K. 2012. Evolutionary Biology, Cell-Cell Communication and Complex Disease. Wiley, Hoboken.

Torday, J.S. and Miller, W.B. 2016. Phenotype as Agent for Epigenetic Inheritance. Biology (Basel) 5: 30.

Torday, J.S. and Rehan, V.K. 2017. Evolution, The Logic of Biology. Wiley, Hoboken.

Torday, J.S. 2018. Pleiotropy, the physiologic basis for biologic fields. Prog. Biophys. Mol. Biol. 136: 37–39.

Chapter 8
Life is a Simulacrum of Cosmologic Physics, Mathematics, Art, Music

Ho: Life is the simulation of physics, mathematics and art, embodied in the Cosmos

Introduction

If we are a simulacrum, derivative of the physical world, what are we an imitation of? In further thinking about the relationship of math to biology, perhaps we are literally the embodiment of the mathematics. So for example, we are taught that the cell is a sphere because that is the optimal configuration for the surface area-to-volume ratio, but that's reasoning after the fact. Perhaps the cell is a sphere because like Joseph Bekenstein's realization that the total energy of a Black Hole is its surface area, not its volume, its event horizon is best expressed as a sphere, its energy being localized to the surface. Therefore, I would propose that the cell is also seen as a surface phenomenon from the perspective of the observer (the Cosmos), the cell having 'invented' itself by assembling lipid molecules as the micelle in equal and opposite reaction to the force of gravity. The oxygenation property is in service to that, not the other way around?

There appears to be an integral relationship between the cell and mathematics, such as mathematical knots as circles, Peter Rowlands' Rewrite Math, Jacob Bekenstein's Black Holes and the cell as holograph, and evolution as Fibonacci numbers, the Golden Ratio as the interrelationship between the Fibonnaci numbers being mathematically equivalent with gravitational force. All of these relationships will be addressed below, raising the question as to causation or association? Suffice to say that the cell is a fractal, formed at the interface between Symbiogenesis and Quantum

Entanglement. Given that, it should, in retrospect, not be a surprise that all of these properties work together efficiently and effectively.

Logically, we tend to think of the mathematics being innate to our consciousness, but as Samuel Butler has said: "A hen is only an egg's way of making another egg", opening up to the possibility that we are merely a simulacrum of the Cosmic mathematics. If nothing else, that way of thinking would answer the perennial question as to why we feel there is something greater than ourselves within us. So here's a reification of this idea.

Homologies between math and the cell

Knot math

Knot Mathematics, for example, informs us that the proof of a true knot is that you can untie it to form a circle. Biologically, a two-dimensional cell is a circle, so is there a homology between them? There may be in the sense that over the course of the life cycle cells 'knot' themselves together into physiologic structures that exhibit cellular cooperativity in support of metabolism. And when cells prepare to reproduce, they unknot themselves in the process of meiosis, reforming as a single germ cell, either egg or sperm. Both cell division, or mitosis, and reduction division, or meiosis, are nominally essential for growth and differentiation of cells. But perhaps there is a bigger picture purpose for these processes in service to evolution. We now know that meiosis is the means by which epigenetic inheritance occurs, the organism assimilating epigenetic 'marks' in the egg or sperm in order to inform the offspring of changes in the environment that it must adapt to when it emerges (Ben Maamar et al., 2021).

Similarly, through the mechanism of internal 'evolution', when cells that form physiologic structures and functions are stressed, the shear force on the micro-vessel walls produce Radical Oxygen Species that are known to cause cellular mutations and duplications (Storr et al., 2013); the net result of such self-engineering to regain homeostasis is what we commonly refer to as evolution (Torday and Rehan, 2012). But the point is that like meiosis, mitosis also serves in the process of evolution.

Rewrite math and the cell

Peter Rowlands' Rewrite Math is homologous with cellular evolution in two fundamental ways. First, Rowlands emphasizes the significance of 'zero' as the attractor function, providing the reference point for the data set. Similarly, the cell acts as the reference point for the evolution of complicated physiologic structure and function (Torday, 2015). Furthermore, the introduction of a new datum into the dataset requires an assessment of its 'fit' with the entirety of the dataset; this is homologous with the way in which the germ cell evaluates epigenetic data during meiosis (Ben et al., 2021).

Bekenstein's black hole and the cell

In his calculation of the total energy of a Black Hole, Jakob Bekenstein realized that it was not a function of the volume of the Black Hole as one might expect, but its surface area as an Event Horizon. That realization prompted Bekenstein to hypothesize that a Black Hole is actually a holograph (Bekenstein, 2003). Similarly, the cell is also an Event Horizon, seen from the outside as a holographic representation due to its negentropic insides (Schrodinger, 1944).

Fibonacci numbers reflect the process of cellular evolution

Each Fibonacci number is the sum of the previous two numbers. This is very much like the way in which cell-cell signaling has responded to environmental factors that have posed an existential threat, the alternating 'leap frogging' of genes and environment being reflected in their responses to global changes in the chemistry and physics of the Earth (Torday and Rehan, 2007).

The fractional mathematical relationship between the Fibonacci numbers is referred to as the Golden Ratio (Livio, 2002). Interestingly, the Golden Ratio is mathematically equivalent to that of gravity (https://www.youtube.com/watch?v=JCvQnnp2FFI), begging the question as to the relationship of the Fibonacci numbers to gravity as the force that generated the protocell (Torday, 2003). Are these properties causal or merely associative ? Given that there was no 'inside' or 'outside' prior to the formation of the cell, it would have to be concluded that the Fibonacci numbers are innate to our physiology, but that they refer to the Fibonacci numbers in the Cosmos, the Golden Ratio being the "imagination" behind the effect of gravity in forming the protocell, as expressed in "Life is a Mobius Strip" (Torday, 2021).

In other words, due to the process of Symbiogenesis as the endogenization and assimilation of existential threats to life as evolution, the mathematics was also acquired during this process as the local reference to the non-local mathematics of the Cosmos. This concept was expressed by Husserl in his explanation for the origin of Geometry in the Cosmos, not being innate to us as we tend to think (Husserl, 1939). That idea is similar to that of Plato's Cave.

Why the relationship between the cell and mathematics?

In order to understand the interrelationship between the math and the physiology, we need to recap cellular evolution from its origin in the protocell. Lipids produced by Pulsars (Smolin, 1997) are ubiquitous in space; when they were immersed in the waters of the primordial ocean that covered the Earth, generated by snowball-like asteroids pelting the planet's surface, they aligned perpendicularly to the surface due to the force of gravity generated by the planet, their negatively-charged poles facing downward due to miscibility, their positively-charged poles pointing upward; when they packed tightly enough together, their negative charge had enough force to dissipate the surface tension due to Van der Waal's force, allowing the lipid molecules to spontaneously form micelles-cell-like spheres with a semi-permeable membrane. The force of gravity impinging on that curved surface produced the internal energy to sustain and perpetuate the Quantum Entanglement of particles within the interior milieu (milieu interieur, Claude Bernard) of the cell. It is that process that gave rise to Symbiogenesis as the basis for all of evolution (Chapman and Margulis, 1998).

Reprising cellular evolution

To understand the interrelationship between the cell and mathematics, you need to go back to the origin of life. Initially, the Earth was covered by water delivered by snowball-like asteroids (Deamer, 2017); lipids are amphiphiles with negatively- and positively-charged ends produced by Pulsars that fell on the water's surface. Those lipid molecules were drawn downward by gravity, their negatively-charged ends pointing downward, their positively-charged ends pointing skyward. Once enough lipid molecules packed together, their negative charge neutralized the Van der Waal's force that produces surface tension, allowing the lipids to coalesce as micelles, or spheres with semi-permeable membranes. Moreover, the gravitational force impinging on the curved surface of the micelle produced energy, which would have 'fueled' Quantum Entanglement of the particles within the micelle, referencing the non-local gravity of the Cosmos.

The partitioning of the environment into 'outside' and 'inside' by the lipid membrane, divided the Implicate Order from the Explicate Order. The assimilation of factors in the environment that posed existential threats, or Symbiogenesis, is the origin of our physiology, giving rise to consciousness as the aggregate of the relationship of our insides and the Cosmologic outsides.

Our communication ability derives from that foundation, whereas mathematics remains in the Cosmos, our physiology making reference to it. This is the basis for the above-cited interrelationships between Knots, Rewrite, Fibonacci, Golden Ratio and physiology. Under conditions of stress like Near Death Experiences, Out of Body Experiences, Maslow Peak Experiences, the Runner's High bring our physiology closer to the math.

The artist's muse as non-locality

Artists speak of their 'muse' as their inspiration to create works of art. The muse is the coalescing of their consciousness of something greater than themselves. In "Life is a Mobius Strip" the concept of a dual track for consciousness, the Explicate as the binary of the cell membrane, and the Implicate as the mapping of the Cosmos onto our physiology acting as our awareness of our surroundings, both the local and the non-local. It is this underpinning of our consciousness that makes us feel that there is something greater than ourselves, and in the case of the artist, that is the drive to express it in painting, sculpture, music, drama, dance. How we transcend the local and approximate the Quantum Mechanical non-local is what the artist is trying to convey to the observer. More often than not, such experiences are precipitated by some stressful event, though it can be learned and habituated if we are open to it. That should be the goal of education, ideally, because the latter is the process of conveying what we know to our offspring in order to avoid extinction.

Being 'in the zone' in sports

In sports, athletes experience being 'in the zone', which is akin to Near Death Experiences, Out of Body Experiences, Maslow Peak Experiences, the Runner's High, and meditation. All of these are manifestations of stress under control by our endocrine system, adrenalin poising our physiology for 'fight or flight'. Adrenalin is essential for both learning and fear, depending upon how much is produced. And in fact learning can modify the 'flight' mode, providing problem solving as a third option, particularly in a social system that protects us from threat of death and fear.

Local, non-local, and synchronicity

Jung was fixated on synchronicity as non-causal coincidence. Is that a consequence of aligning our local physiology with the non-locality of the Cosmic references of math, physics, art? Is the article "Life is a Mobius Strip" it was conjectured that Symbiogenesis emerged from Quantum Entanglement, forming a continuum from the fundament of physics to complex physiology, and everything in between. With that lens, why not think we could align with our non-local reference points in order to transcend the locality of our physiology? Is this what we do when we meditate?

In reality, the Scientific Method is a form of synchronicity because in order for the scientific community to acknowledge the validity of a scientific observation, it must be reproduced by another scientist. It is that principle of independent observation that coincides with Jungian synchronicity. It is this way of understanding synchronicity which takes it out of the realm of metaphysics and centers it in 'evidence-based information', the gold standard of contemporary knowledge.

Science, Art and C.P. Snow's "Two Cultures" …..No More

In his 1959 essay, C.P. Snow lamented the segregation of science and the arts (Snow, 1959). That separation has lasted to this day because we fail to recognize the nature of our consciousness as the aggregate of our physiology (Torday, 2018), which references the non-local source of these perceptions in the Cosmos. They have been assimilated over the course of evolution, and dictate our behavior as 'agents', collecting epigenetic data over the course of our lifespan in order to inform the next generation of on-coming threats to our existence. By incorporating such epigenetic 'marks' in the DNA of our germ cells (egg and sperm), the offspring is pre-tuned to existential threats.

Archimedes, Proust, Flemming, Ignarro

Rest assured, Archimedes was not the first person to get out of a bathtub, Proust was not the first to bite into a madeleine, Flemming was not the first to notice yeast contamination of a bacterial culture, and Ignarro was probably not the first to smoke while studying a muscle strip preparation connected to a polygraph…..but each was in the proper 'alignment' to gain insight to the relationship between those local experiences and the non-local references discussed above. "In the field of observation, chance favors only the prepared mind" (Louis Pasteur).

Local v non-local as the quantum mechanics of math, physics, art as a TOE?

All of our conscious activities—mathematics, physics, art—have both local and non-local aspects. As expressed in "Life is a Mobius Strip", that dualism is due to the Explicate Order as the everyday functioning of the cell membrane as the determinant of whether any given factor should be inside or outside of the cell; the Implicate Order connects our physiology to the Cosmos as the aggregate of the Symbiotic process by which we

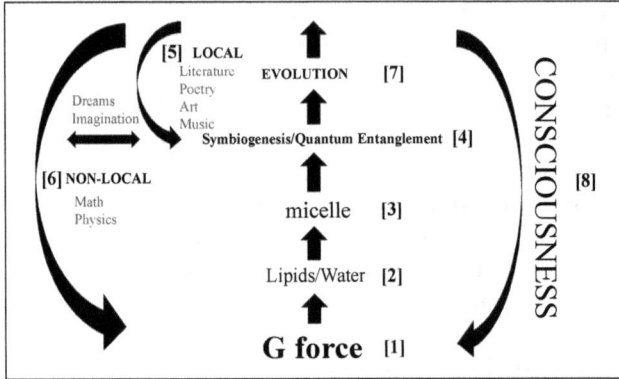

Figure. Evolution by Imitation. [1] Gravity force on [2] lipids in water [3] gave rise to micelles. Gravity impinging on micelles produced the energy necessary for [4] Quantum Entanglement as the basis for Symbiogenesis. [5] Such local properties [6] reference non-local Cosmology [8] over the course of Evolution [7]. The aggregate [1–7] is Consciousness [8].

have assimilated existential threats in the environment such as bacteria, heavy metals and gases [see Figure]. The compartmentation and linking of such factors through cell-cell communication with soluble growth factors and their receptors is governed by the principle of Terminal Addition, or the adding on of newly acquired traits at the end of a series of evolved traits (Torday and Miller, 2018). That feature provides the complementarity between the Cosmos and our physiology as both non-local and local, respectively, expressed as Near Death Experiences, Out of Body Experiences, the Runner's High, Maslow Peak Experiences and meditation. And perhaps those experiences are represented by 'Receiver Operating Characteristic Curves', distinguishing 'true' from 'false' positives, particularly as it applies to radar. That interrelationship would also provide the explanation for the local and non-local characteristics of math, physics and art.

Discussion

Key to formulating a true narrative for our ontology and epistemology is an understanding of how and why we have evolved from the physical. That process was elucidated by the Margulis theory of Symbiogenesis, centered initially on how bacteria were assimilated as mitochondria, for which there is now more than ample evidence. But the theory goes far beyond just mitochondria originating from the assimilation of bacteria, essentially being accounted for by every step in the process of evolution as the basis for emergent traits (Torday and Rehan, 2017).

Once seen in the forward direction as serial solutions to existential threats based on cell-cell communication, the origin and course of evolution can be seen as causal, not random as Darwin would have us think. The first empiric evidence for this was recently published, showing that the response to malarial infection was not a random mutational event (Melamed et al., 2022). However, it did not offer the mechanism underlying the non-random inheritance mechanism. Yet in a long series of journal articles and books, Torday et al. have hypothesized that the formation of the cell membrane from lipid molecules as micelles is the origin of life in response to the force of gravity (Torday, 2021). That concept directly links non-life with life, and the residuals of that can still be detected in the mechanism of action of Parathyroid Hormone-related Protein, a 'stretch-regulated' gene that was necessary for the transition from water to land. Experimentally, if differentiated lung or bone cells are subjected to microgravity they lose their identity (Torday, 2003), providing direct evidence for the role of gravity in the initiation of life. In further support of this concept, it was hypothesized that symbiogenesis as the basis for evolution emerged as a consequence of Quantum Entanglement, locally sustained and perpetuated by the force of gravity impinging on the curved surface of the protocell producing energy (Einstein's Field Theory), referencing the non-local force of gravity in the Cosmos. That energy source would have facilitated the Quantum Entanglement of particles passing through the semi-permeable cell membrane (Torday, 2021). Thus the long-awaited link between Quantum Mechanics and biology has been discovered. More importantly, the iterative assimilation of existential threats in the environment were mediated by symbiogenesis, providing the origin of our physiology. The net result is that our physiology is the source of our consciousness as our awareness of the Cosmos, which has mapped onto our physiology (Torday, 2020). The intermediate properties of our physiology as cell-cell communication (Torday and Rehan, 2012) and Terminal Addition (Torday and Miller, 2018) are hallmarks of the relationship of our physiology and consciousness, though up until now they have been seen as anecdotal, not causal. Yet it is the cell-cell communication that has mediated the flow of energy through our physiology, and provided the cuing for injury repair mechanisms for restoration of physiologic homeostasis (Demayo et al., 2002). Terminal Addition, or the practice of adding newly acquired traits onto an evolved series of traits, is a direct consequence of cell-cell communication mediated by soluble growth factors, obviating other than terminal addition because to do otherwise would place the prevailing cell-cell communication principle at risk of extinction.

Needless to say, the above-cited novel way of perceiving our origin and method of survival offers a way of understanding the nature of our being holistically for the first time.

References cited

Bekenstein, J.D. 2003. Information in the Holographic Universe—Theoretical results about black holes suggest that the universe could be like a gigantic hologram. Scientific American 289: 58–65.

Ben Maamar, M., Nilsson, E.E. and Skinner, M.K. 2021. Epigenetic transgenerational inheritance, gametogenesis and germline development†. Biol. Reprod. 105: 570–592.

Chapman, M.J. and Margulis, L. 1998. Morphogenesis by symbiogenesis. Int. Microbiol. 1: 319–326.

Deamer, D. 2017. The role of lipid membranes in life's origin. Life (Basel) 7: 5.

Demayo, F., Minoo, P., Plopper, C.G., Schuger, L., Shannon, J. and Torday, J.S. 2003. Mesenchymal-epithelial interactions in lung development and repair: are modeling and remodeling the same process? Am. J. Physiol. Lung Cell Mol. Physiol. 283: L510–L517.

Husserl, E. 1939. The origin of geometry. Rev. Int. Philos. 1: 157–179.

Livio, M. 2002. The Golden Ratio. Broadway Books, New York.

Melamed, D., Nov, Y., Malik, A., Yakass, M.B., Bolotin, E., Shemer, R., Hiadzi, E.K., Skorecki, K.L. and Livnat, A. 2022. *De novo* mutation rates at the single-mutation resolution in a human HBB gene-region associated with adaptation and genetic disease. Genome Res. 14: gr.276103.121.

Schrodinger, E. 1944. What is Life? MacMillan, New York.

Smolin, L. 1997. The Life of the Cosmos. Oxford University Press, Oxford.

Snow, C.P. 1959. Two Cultures. Science 130: 419.

Storr, S.J., Woolston, C.M., Zhang, Y. and Martin, S.G. 2013. Redox environment, free radical, and oxidative DNA damage. Antioxidant Redox Signal. 18: 2399–2408.

Torday, J.S. 2003. Parathyroid hormone-related protein is a gravisensor in lung and bone cell biology. Adv. Space Res. 32: 1569–1576.

Torday, J.S. and Rehan, V.K. 2007. Developmental cell/molecular biologic approach to the etiology and treatment of bronchopulmonary dysplasia. Pediatr. Res. 62: 2–7.

Torday, J.S. and Rehan, V.K. 2012. Evolutionary Biology, Cell-Cell Communication and Complex Disease. Wiley, Hoboken.

Torday, J.S. 2015. The cell as the mechanistic basis for evolution. Wiley Interdiscip. Rev. Syst. Biol. Med. 7: 275–284.

Torday, J.S. and Rehan, V.K. 2017. Evolution, the Logic of Biology. Wiley, Hoboken.

Torday, J.S. 2018. From cholesterol to consciousness. Prog. Biophys. Mol. Biol. 132: 52–56.

Torday, J.S. and Miller, W.B. Jr. 2018. Terminal addition in a cellular world. Prog. Biophys. Mol. Biol. 135: 1–10.

Torday, J.S. 2020. Consciousness, Redux. Med. Hypotheses 140: 109674.

Torday, J.S. 2021. Life is a mobius strip. Prog. Biophys. Mol. Biol. 167: 41–45.

Chapter 9
Morphological Forms Emerging from the Evolutionary Process are Topologies

Introduction

Evolution has proven refractory to experimentation—how fish have evolved into hominins phylogenetically, for example. That is basically because evolution is purported to be due to random mutations, which precludes the possibility of testing such a mechanism, leaving only correlations and associations, which are not causal. Some comparable causal process that could be exploited in order to deconvolute evolution would be helpful, but none was forthcoming….up until now, having been introduced to Knot Theory (Kauffman, 2006), which is a subset of Topology. Topology is the mathematical expression for how geometric objects are able to maintain themselves under continuous deformations. In turn, Knot Theory is the study of mathematical knots, inspired by the knots that occur in everyday life. Since knots tie things together, they are conceptually comparable with the cell-cell signaling mechanisms that tie physiologic traits together during development, like the mesenchymal-epithelial interconnections that form phenotypic structure and function in various organs. Such interactions terminate in physiologic homeostasis.

Mathematical knots are held together conceptually by theoretical 'springs' (Kauffman, 2006), and can be untied by freeing them from their basement membranes due to the specific actions of metaloproteases (Sternlicht and Werb, 2001), re-forming as circles. In the case of physiologic knots, they untie themselves during the process of meiosis, under physiologic stress, and during the process of aging or senescence, ultimately recapitulating the unicellular state as a 'circle' at the end of the life cycle.

The homology between knots and evolution

Knots are surfaces, which is how they relate to evolution. Lipids originating either from the Cosmos (Salama, 2008) or from thermal vents in the sea floor localized to the air-water interface [Figure 1], where they formed micelles, prototypical cells that produced a surface between the Explicate and Implicate Orders, forming what Claude Bernard referred to as the milieu interieur. This is the origin of unicellular organisms, which dominated the Earth for the first 3.5 billion years. Due to competition between prokaryotes and eukaryotes, eukaryotes began forming multicellular organisms through cell-cell communications mediated by growth factor-growth factor receptor signaling mechanisms. Such cellular-molecular pathways for growth and differentiation during embryologic development subsequently became the homeostatic basis for physiology (Torday and Rehan, 2017), as mentioned above. In doing so, the cells involved produce extracellular matrices that stabilize them physically (Sottile and Hocking, 2002), forming real 'knots' that literally tie physiology together as organisms. During the etiology of chronic disease, such cellular communications break down, reverting back to earlier developmental and phylogenetic homeostatic stages of evolution, as in emphysema and chronic glomerular disease (Demayo et al., 2002), stabilized by scar tissue (Yates et al., 2011). During the aging and senescence,

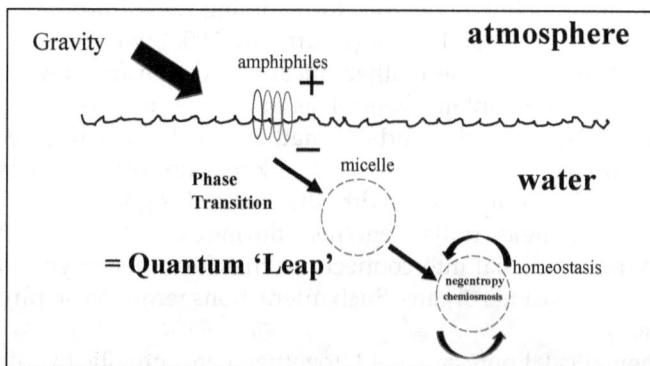

Figure 1. Formation of Micelles from lipids immersed in water. When lipids are immersed in water, they initially align perpendicularly to the surface as amphiphiles, their hydrophobic ends facing upward, their hydrophilic ends facing downward. These lipids will then spontaneously form micelles, spheres with semi-permeable membranes surrounding them.

loss of bioenergy similarly leads to systematic loss of homeostatic control and deterioration of cell-cell signaling (Torday and Rehan, 2012), revealing the knot-like characteristics of organismal structure and function.

And it should be borne in mind that all of the above reduces to Quantum Mechanics, Quantum Entanglement being the homolog of Symbiogenesis that sustains and perpetuates homeostasis.

Over the history of vertebrate evolution, organisms have produced the phylogenetic changes of speciation. But it should be remembered that organisms always return to the unicellular state over the course of their life cycle. The capacity to form and unform structure and function, all the way back to a 'circle' (Kauffman, 2006) is also the property of a mathematical knot.

Circles, knots, cells, memory

Memory is necessary for the process of evolution—in order to remain in sync with an ever-changing environment, the organism must remember its past ways of adapting as the most efficient way of surviving, referred to as exaptations. It is conventionally held that DNA is the biologic form of memory, but the use of nucleic acids for this purpose is derivative of the lipid-based micelle, which has molecular memory in the form of hysteresis. The ability of lipids to deform when warmed by the sun, and re-form in the cold of night may have been the genesis of memory, given that the brains of birds and mammals cool during non-Rapid Eye Movement sleep (Lyamin et al., 2018). Like this most basic of functional properties of life, knots also have memory, because as complicated as they may become as Trefoils, Borromeans, etc., they remain simple circles at their origin as the formal validation that they are in fact knots (Kauffman, 2006). The same holds true for a physiologically complicated organism, because it ultimately returns to its unicellular form over the course of its life cycle, proving that it, too, is a knot. By returning to the unicellular state, the organism has a way of integrating epigenetic 'marks' into its physiology over the course of ontogeny (Ben Maamar et al., 2021), explaining the cellular basis for evolution.

Short of regressing to the unicellular state, tissues can default to earlier stages in their ontogeny and phylogeny through loss of cell-cell signaling (Guex et al., 2020). Guex's model organism for this phenomenon are ammonites, which are of particular interest because the 'chambers' of the Nautilus allow for adaptation to gravity in the same way that the fish swim bladder does (Trueman, 1940), and Parathyroid Hormone-related Protein has been implicated in both structures (Clark et al., 2020), providing the homology between them. This mechanism is due to cellular networking for development and homeostasis, providing the option of reverting to an earlier physiologic state in order to maintain equipoise, allowing the organism to reproduce and transfer its genome to the next generation.

DNA knots, homeobox genes, cell-cell signaling

Evidence for DNA 'knots' being altered by changes in cell-cell signaling has not been forthcoming. However, we know that under various stress conditions, cells produce Radical Oxygen Species that may cause mutations and duplications (Storr et al., 2013). Such DNA changes are acquired by the genome, hypothetically altering the topology of DNA. The best evidence for this is the fact that Homeobox gene DNA is arrayed along chromosomes in the same way that it appears in the organism (Krumlauf, 1992).

Pre-adaptations, lipids immersed in water, and knots

The cellular approach to evolution is predicated on serial pre-adaptations, or exaptations (Gould and Vrba, 1982), raising the question as to where, why and when the knots for life began? In the scenario for the role of lipids in water as the origin of life, lipids on Earth arose from both thermal vents in the sea floor as mentioned earlier, or as hitch-hikers on the asteroids that delivered frozen water to earth, the lipids originating from the Cosmos (Salama, 2008).

At first, the lipids aligned perpendicularly to the water-atmosphere interface [Figure 1] because they are polarized amphiphiles, their positively-charged ends facing upwards, being hydrophobic, their negatively-charged hydrophilic ends immersed in the water. In that configuration they act like an antenna, responding to electromagnetic forces such as Pulsars and photons. When the lipids form micelles, which have coped with their ever-changing environments by assimilating factors in their environment-heavy metals, gases, ions, bacteria—referred to as the Symbiogenesis Theory (Sagan, 1967) [Figure 2]. Once such factors

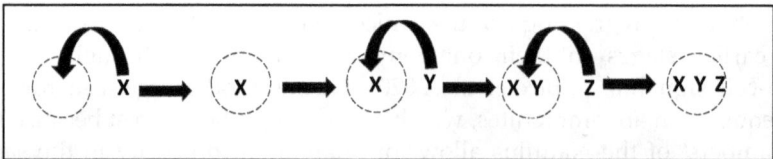

Figure 2. Endosymbiogenesis. Over the course of evolution, cells have endogenized factors in their environment that have posed existential threats, represented by x, y and z. The compartmentalization of such factors is the basis for cellular physiology.

were endogenized, they were compartmentalized using intracellular membranes, forming what is conventionally referred to as cellular physiology. Organisms became progressively more complicated based on cellular communications (Torday and Rehan, 2012), are spheres with semi-permeable membranes, the effect of the electromagnetic force acts to stimulate cytoplasmic streaming as a phenotypic flow of energy. Subsequently, the appearance of cholesterol in the cell membrane caused thinning of the membrane, further facilitating cytoplasmic streaming as locomotion. Fast forward to Ciona intestinalis, a unicellular organism in which the stem cells of the heart develop in the tail, and then migrate into the body to form the heart (Davidson and Levine, 2003). It is within the realm of possibility to think that the beating of the tail morphs into the beating of the heart. This process may have exaptively evolved from the earlier evolution of cytoplasmic streaming since the heart evolved as a muscularization of the vasculature, like the heart of an earthworm, perhaps due to the homologous 'cytoplasmic streaming' of blood causing shear stress, known to generate gene mutations and duplications by generating Radical Oxygen Species [Figure 3]. The one-chambered heart of the earthworm could have given rise to the two-chambered heart of fish, the three-chambered heart of amphibians and reptiles, and ultimately to the four-chambered hearts of birds and mammals due to beta-adrenergic

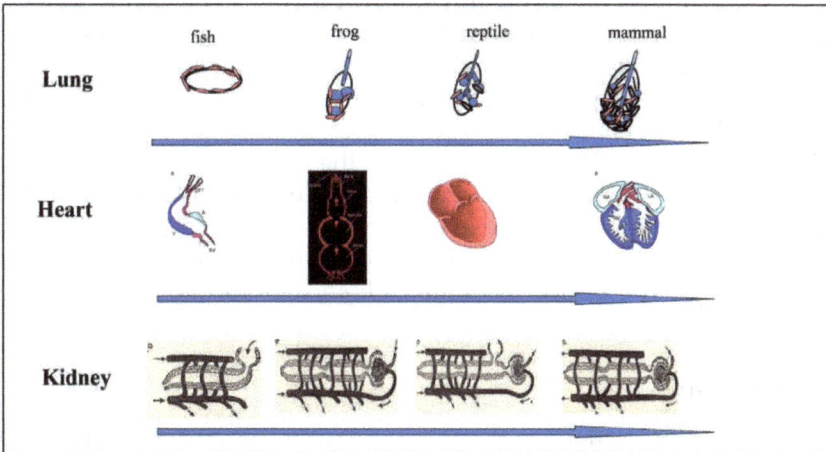

Figure 3. Evolution of the lung, the heart and the kidney. The evolution of the lung (top), heart (middle) and kidney (bottom) are depicted developmentally and phylogenetically. Note the increasing complexity of these phenotypic traits from left to right as a result of the interaction of a changing environment and the cellular components of each structure/function based on cell-cell signaling controlled by homeostasis.

agents like adrenaline. If you experimentally delete the beta-adrenergic receptor from developing mice, for example, they are born with a two-chambered heart. The beta-adrenergic receptor 'duplicated', i.e., amplified during the water-land transition (Aris-Brosou et al., 2009), likely promoting the progressive evolution of the heart from two chambers in fish, to three chambers in amphibians and reptiles, to four chambers in birds and mammals.

The above provides a 'knot' that begins with the lipids forming a boundary between the Explicate and Implicate Orders, all the way through to the evolution of the lung, heart and kidney. The genetics of heart development, for example, are connected to those of the fingers, exemplifying the phenotypic homology between these structures based on cytoplasmic streaming.

The deep significance of the homology between knots and evolution

The mechanism of cellular evolution is predicated on exaptations from sequentially earlier and earlier genetic traits, working backwards from present-day physiology to the unicellular state. Beginning with lung evolution proved to be fortuitous because gas exchange is a very ancient property of life that allowed the tracing of the process all the way back to the unicell (Torday, 2018). But that begs the question as to what the prototype for the cell was? The only such unity seems to be the Singularity of nature that existed prior to the Big Bang (Hawking, 1998). The rationale is that the unicell has the capacity to express all of the states of being—homeostasis, mitosis and meiosis—all three of which are determined by the cytoskeleton as the 'sensor' for the prevailing environmental conditions. But the insight that knots and evolution share common properties suggests that even before the homology between the cell and the Singularity, the shared topology of knots and biology may have formed the ultimate homology for the inanimate and animate.

Discussion

Random gene mutations as the cause of Darwinian evolution dissociate that process from development and phylogeny as sequential processes. Conversely, the reverse-engineering of cell-cell communications has provided fundamental insights to the 'how and why' of the ontology and epistemology of evolution. The current delineation of the interrelationship between cell-cell communication and Knot Theory furthers such fundamental causal mechanisms, providing support for the previously expressed concept of The Singularity of Nature (Torday, 2019). It is only through such cross-cutting, transcendent diachronies that we will ultimately be able to form an algorithmic perspective for a holism of life in the Cosmos.

Up till now it has only been the artists who have been able to ex- press the holism of our existence. Robert Frost said that 'life is that which can mix oil and water', and that 'The Secret Sits' in the middle, and we dance around wondering why. And in Little Gidding, T.S. Eliot assures us that

"We shall not cease from exploration

And the end of all our exploring
Will be to arrive where we started

And know the place for the first time."

In art, Magritte's painting "Not to be Reproduced" [Figure 4] depicting a man seeing himself in a mirror from behind expresses the observer and the observed, while Henry Moore confronts us with the question as to whether the negative space in his sculptures has the same or greater value than the granite or bronze? The merging of evolution and topology offers

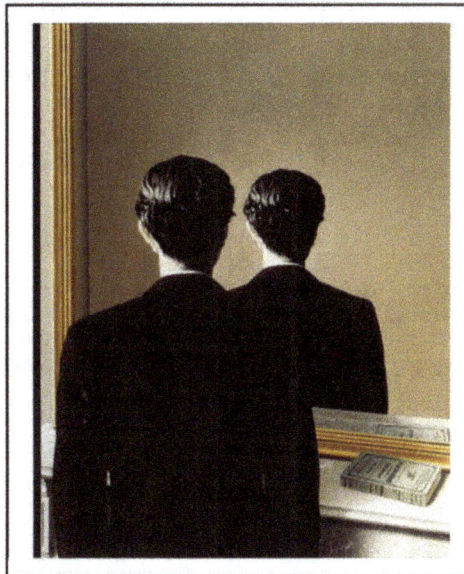

Figure 4. Magritte's "Not to be reproduced".

the opportunity to solve these enigmas scientifically and mathematically at long last.

References cited

Aris-Brosou, S., Chen, X., Perry, S.F. and Moon, T.W. 2009. Timing of the functional diversification of alpha- and beta-adrenoceptors in fish and other vertebrates. Ann. N. Y. Acad. Sci. 1163: 343–347.

Ben Maamar, M., Nilsson, E.E. and Skinner, M.K. 2021. Epigenetic transgenerational inheritance, gametogenesis and germline development. Biol. Reprod. 105: 570–592.

Clark, M.S., Peck, L.S., Arivalagan, J., Backeljau, T., Berland, S., Cardoso, J.C.R., Caurcel, C., Chapelle, G., De Noia, M., Dupont, S., Gharbi, K., Hoffman, J.I., Last, K.S., Marie, A., Melzner, F., Michalek, K., Morris, J., Power, D.M., Ramesh, K., Sanders, T., Sillanpää, K., Sleight, V.A., Stewart-Sinclair, P.J., Sundell, K., Telesca, L., Vendrami, D.L.J., Ventura, A., Wilding, T.A., Yarra, T. and Harper, E.M. 2020. Deciphering mollusc shell production: the roles of genetic mechanisms through to ecology, aquaculture and biomimetics. Biol. Rev. Camb. Phil. Soc. 95: 1812–1837.

Davidson, B. and Levine, M. 2003. Evolutionary origins of the vertebrate heart: specification of the cardiac lineage in Ciona intestinalis. Proc. Natl. Acad. Sci. U.S.A. 100: 11469–11473.

Demayo, F., Minoo, P., Plopper, C.G., Schuger, L., Shannon, J. and Torday, J.S. 2002. Mesenchymal-epithelial interactions in lung development and repair: are modeling and remodeling the same process? Am. J. Physiol. Lung Cell Mol. Physiol. 283: L510–L517.

Gould, S.J. and Vrba, E.S. 1982. Exaptation, a missing term in the science of form. Paleobiology 8: 4–15.

Guex, J., Torday, J.S. and Miller Jr., W.B. 2020. Morphogenesis, Environmental Stress and Reverse Evolution. Springer, Cham, Switzerland.

Hawking, S. 1998. A Brief History of the Universe. Bantam, New York, New York.

Kauffman, L. 2006. Formal Knot Theory. Dover Books, Mineola, NY.

Krumlauf, R. 1992. Evolution of the vertebrate Hox homeobox genes. Bioessays 14: 245–252.

Lyamin, O.I., Kosenko, P.O., Korneva, S.M., Vyssotski, A.L., Mukhametov, L.M. and Siegel, J.M. 2018. Fur seals suppress REM sleep for very long periods without subsequent rebound. Curr. Biol. 28: 2000–2005.

Sagan, L. 1967. On the origin of mitosing cells. J. Theor. Biol. 14: 255–274.

Salama, F. 2008. PAH's in Astronomy—a Review. Organic Matter in Space. Proceedings I.A.U. Symposium 251: 357–365.

Sottile, J. and Hocking, D.C. 2002. Fibronectin polymerization regulates the composition and stability of extracellular matrix fibrils and cell-matrix adhesions. Mol. Biol. Cell 13: 3546–3559.

Sternlicht, M.D. and Werb, Z. 2001. How matrix metalloproteinases regulate cell behavior. Annu. Rev. Cell Dev. Biol. 17: 463–516.

Storr, S.J., Woolston, C.M., Zhang, Y. and Martin, S.G. 2013. Redox environment, free radical, and oxidative DNA damage. Antioxidants Redox Signal. 18: 2399–2408.

Torday, J.S. and Rehan, V.K. 2012. Evolutionary Biology, Cell-Cell Communication and Complex Disease. Wiley, Hoboken, NJ.

Torday, J.S. and Rehan, V.K. 2017. Evolution, the Logic of Biology. Wiley, Hoboken, NJ.

Torday, J.S. 2018. A diachronic evolutionary biologic perspective: reconsidering the role of the eukaryotic unicell offers a 'Timeless' biology. Prog. Biophys. Mol. Biol. 140: 103–106.

Torday, J.S. 2019. The Singularity of nature. Prog. Biophys. Mol. Biol. 142: 23–31.

Trueman, A.E. 1940. The ammonite body-chamber, with special reference to the buoyancy and mode of Life of the living ammonite. Q. J. Geol. Soc. Lond. 96: 339–383.

Yates, C.C., Bodnar, R. and Wells, A. 2011. Matrix control of scarring. Cell. Mol. Life Sci. 68: 1871–1881.

Chapter 10
On the Quantum Origin and Nature of Consciousness

Introduction

The origin and nature of consciousness remain undetermined, yet the consensus is that it is the most important thing we need to understand (Koch, 2018). Perhaps it is because it is so fundamental that it is like air or terra firma, and is integral to our very being, rendering it inaccessible using conventional logic and analogy. It is complicated by the fact that as we have evolved from the 'ambiguity' of negative entropy (Schrodinger, 1944; Torday and Miller, 2017), we have defied gravity by causing the twists and turns of embryologic development (Levin, 2005), reinforced by deuterostomy, developing from anus to mouth. That process of negative gravitropism codified by the course of the vagal nerve, the largest nerve in the autonomic nervous system, emanating from around the adrenals, passing up through the gut to the heart, and face, influencing consciousness (Porges, 2007). As Proof of Principle, it is hypothesized that the left-right brain rectifies the 'backwardization' of our orientation to gravitational force, acting to resolve the Newtonian properties as Quantum Mechanical (McGilchrist, 2009).

The confusion generated by the above has been rectified by the astronomical observation of the 'redshift' (Hubble, 1929), providing experimental evidence for a rational testable and refutable beginning for existence. It is that finding that has given impetus to finding the origin of life and its means of evolving as tractable. And in fact, it is that series of events that has provided clues to the emergence of life, buoyed by molecular level knowledge of development and phylogeny, along with mechanisms of injury-repair (Demayo et al., 2002).

The reverse-engineering of physiology based on cell-cell communication

The first organ to be reverse-engineered in this way was the lung (Torday and Rehan, 2007), for which extensive molecular knowledge of its development, phylogeny and injury-repair are fortunately known [Figure 1]. Developmentally, it is an outpouching of the trachea as a homologue of the swim bladder in Physostomus boney fish, having a tube that connects the swim bladder to the esophagus, called the pneumatic duct, i.e., a trachea; in contrast to that, Physoclistous boney fish such as the Zebra Fish have a swim bladder that absorbs gases directly from the circulation (Zheng et al., 2011). It is known that fish made at least five independent attempts to evolve from water to land about 500 million years ago based on fossil evidence (Clack, 2012). The adaptation to gravity using buoyancy was key to understanding the evolution of the lung, underpinned by genetic signaling mechanisms common to both organs (Torday, 2003). The cellular-molecular mechanism for gravitational adaptation in the lung is Parathyroid Hormone-related Hormone (PTHrP), a stretch-regulated gene that is necessary for the formation of alveoli in the lung (Torday, 2003). The phenotypic link between the swim bladder and lung surfactant, a soapy material that prevent the swim bladder walls from sticking to one another, whereas in the alveoli, surfactant was essential for the progressive decrease in alveolar diameter in order to maximize gas-exchange and accommodate the ever-increasing demand for metabolism on land.

Lipids have played an essential role in evolution, beginning with the formation of micelles as the first cells on earth (Torday and Miller, 2016). They have been particularly important in facilitating oxygenation,

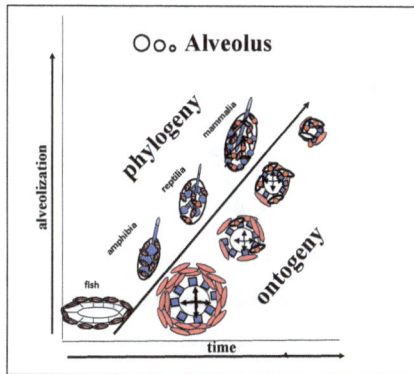

Figure 1. Ontogeny and Phylogeny of the Lung. The ontogeny and phylogeny of the lung are depicted as alveolarization (vertical axis) regressed against time. The progressive decrease in alveolar diameter is depicted at the top of the schematic.

both in their role in alveolar evolution (see above) and as natural anti-oxidants (Torday et al., 2001). It is that consistent relationship that came to the fore when considering the advent of cholesterol synthesis (Torday, 2018a). Konrad Bloch had hypothesized that cholesterol was a molecular fossil because it takes 11 atoms of oxygen to synthesize one molecule of cholesterol (Bloch, 1992). However, he was reasoning after the fact because lipids were essential for the formation of the cell in the first place, so the onset of cholesterol synthesis emerged due to the rise in atmospheric oxygen during the Phanerozoic Era (Berner, 1999). That provided insight to the role of lipids in the formation of micelles from lipid molecules that originated in Pulsars in deep space hitchhiking on snowball-like asteroids that pelted the earth, forming the ocean that covered the earth (Deamer, 2017).

The evolution of consciousness

The lipid molecules floating on the surface of the ocean were drawn downward by the force of gravity [see Figure 2], their negatively-charged ends pointing downward into the water because they are hydrophilic. When such lipid molecules reached a critical mass their negative charge neutralized the Van der Waal's force for surface tension, allowing them to make a 'quantum leap' springing into forming micelles. Such prototypical 'cells' floating on the surface of the ocean would be warmed by the sun by day, causing them to deform, reforming at night when they cooled since lipids exhibit hysteresis, or molecular memory [Figure 3]. The iterative expansion and contraction of the micelles caused them to accumulate calcium ions, which will denature lipids in high enough concentrations;

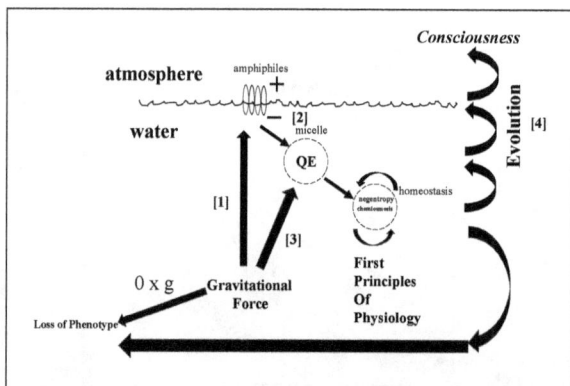

Figure 2. Continuum from lipids to evolution. [1] gravity aligns lipid molecules at the atmosphere-water interface, producing micelles [2]; gravity impinges on micelles [3], providing the energy to sustain Quantum Entanglement of particles within the micelles, leading to the First Principles of Physiology as the basis [4] for Evolution.

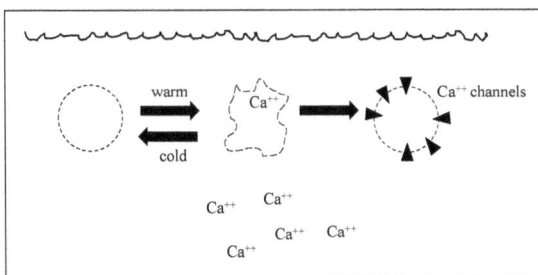

Figure 3. Expansion and contraction of micelles. The warming and cooling of micelles floating on the surface of the ocean caused the concentration of calcium within them. Our forebears evolved channels to control the amount of calcium within the cell.

our forebears evolved calcium ion channels (Case et al., 2007), which evolved into neuronal memory (Gasque, 2015).

Subsequently, [see Chapter 9, Figure 2] Symbiogenesis (Sagan, 1967) was the means by which organisms coped with existential threats, assimilating them as physiology (Torday and Rehan, 2012). But since there is a logic in the Cosmos that stems from stars producing light through stellar nucleosynthesis (Smolin, 1999), giving rise to the Elements in the order of their atomic mass. When organisms assimilated the Elements, they acquired that logic. It is that relationship between star-derived Elements and life that we refer to as consciousness.

As Proof of Principle, the thyroid gland has evolved for the organification of iodine, an otherwise toxic chemical. The first 36 elements are formed by stellar nucleosynthesis in the same order as their atomic mass, whereas Iodine is atomic number 53, leaving a gap in the 'logic' of stellar nucleosynthesis of 17 Elements. The fact that biology 'remembers' how it ascribed to the first formed Elements in the evolution of the thyroid from the endostyle of lampreys is a testament to the role of memory in the process.

But even beyond memory, the connection between Symbiogenesis and Quantum Entanglement addressed in an earlier Chapter provides the ultimate insight to the merging of the animate with the inanimate. Both processes are in service to homeostasis as the way in which life is sustained physiologically and perpetuated via evolution.

Mobius strip motif and imagination

As was said above, micelles gave rise to protocells, which act as probes for epigenetic 'marks' or inhomogeneities in the environment. Such marks are then assimilated in the egg and sperm, where they cause the formation of biochemical adducts that change the readout of DNA, forming adaptive changes that prevent the extinction of the organism. To begin

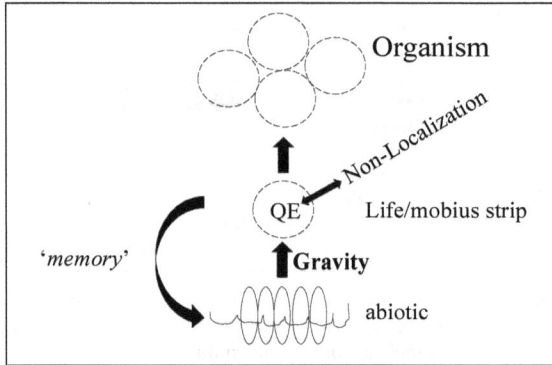

Figure 4. Local and Non-Local Roles of Gravity. Local gravity provided energy for Quantum Entanglement (QE), giving rise to life through Symbiogenesis.

understanding the origin of this mechanism we have to understand the nature of the micelle, delineated by a semipermeable lipid cell membrane that forms a barrier between the inside and outside. But recall that all such milestones in evolution are characterized by emergences, and in the case of the micelle [Figure 4], Symbiogenesis provided the mechanism for evolution, but where did that emerge from? When gravity impinges on a curved surface like that of a micelle, it produces energy. In the case of the evolving cell, that energy was used to promote and sustain Quantum Entanglement of particles passing into and out of the micelle. The gravitational force is local, emanating from the Earth, referencing the gravitational force produced by the Big Bang.

Furthermore, the emergent cell remembers when it wasn't, i.e., when it was lipid molecules. From that point of view, the cell membrane is a functional Mobius Strip, or one imagined continuous membrane without an inside or outside. That trait gives the cell the capacity for imagination.

The nature of consciousness based on its origin

Based on the chain of events leading up to the formation of the unicell, one level of consciousness is the binary decisions made by the lipid cell membrane. The particles allowed into the cell then are assimilated through Symbiogenesis to form physiologic traits [Figure 5], linked together by cell-cell communication. That forms the physiologic continuum for consciousness that references the Cosmos. The interrelationships between the binary of the cell membrane and the continuum of physiology are the constituents of consciousness referencing the Consciousness of the Cosmos. Such events as Near Death, Out of Body experiences, Maslow Peak Moments and the Runner's High constitute wave collapses that bring the two levels of consciousness in closer proximity to one another.

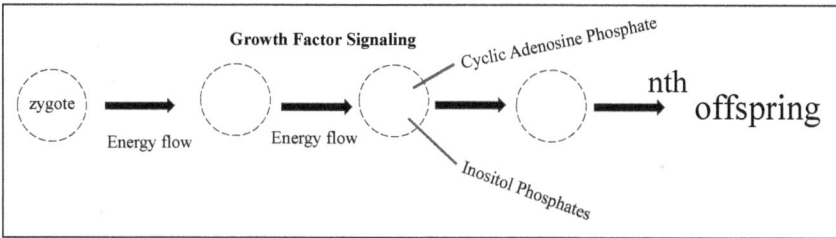

Figure 5. Cell-Cell Signaling via High Energy Flow. Growth Factor/receptor signaling during embryology is mediated by high energy phosphates.

Hypothetically, this kind of phenomenology may be what Lynnclaire Dennis has described as the Merion Matrix, a form of consciousness (McNair et al., 2018).

Role of memory in evolution

The process of evolution is contingent on memory. In order to remain in synchrony with an ever-changing environment, the organism must remember its past ways of adapting as the most efficient way of surviving, referred to as exaptations (Gould and Vrba, 1982). It is widely thought that DNA is our biologic form of memory, but this use of nucleic acids derives from the lipid micelle, which has molecular memory in the form of hysteresis. The focus on DNA as memory is indicative of the misconception that we are material, when in fact we are energy flows (Whitehead, 1979). For example, the process of embryologic development is conventionally seen as cell division giving rise to the various stages of the burgeoning offspring. In fact, it is a series of high energy state changes mediated by soluble growth factors and their cognate receptors (Lodish et al., 2008). And that cascade culminates in physiology as homeostasis, maintaining the balance of energy within the organism (Torday and Rehan, 2012, 2017). When that balance is perturbed, it is sensed as a loss of free energy, and the healing process is the effort to reconstitute homeostasis. Persistent loss of energy results in evolutionary adaptation or extinction (Torday and Rehan, 2012, 2017).

The origin of memory is in the capacity of micelles to deform when warmed by the sun, and re-form in the cold of night. The recursion of that process day after day, night after night results in the accumulation of calcium ions, which can become toxic to lipids. Our forebears evolved calcium ion channels to mediate the entry and exit of calcium (Case et al., 2007).

Like this most basic functional property of life, mathematical knots also have memory, because as complicated as they may become, as Trefoils or Borromeans, they remain simple circles at their origin, exemplified by their formal definition in being able to unknot to form a circle. The same property holds true for a multicellular organism, which ultimately returns to its unicellular form over the course of its life cycle, demonstrating that it too is a knot. By recapitulating the unicellular state, the organism provides a way of integrating epigenetic 'marks' into its physiology over the course of its ontogeny, explaining the cellular basis for evolution.

Short of regressing all the way back to the unicellular state, tissues can default to earlier stages in their ontogeny and phylogeny due to loss of cell-cell signaling (Guex et al., 2020). This mechanism is due to the cellular networking for development and homeostasis, allowing for the option of defaulting to an earlier physiologic state in order to maintain itself, allowing the organism the opportunity to reproduce and transfer its genome to the next generation.

Rowlands Rewrite Math

As independent evidence for the role of mathematics in the process of evolution, the physicist Peter Rowlands, University of Liverpool, has devised a mathematical system he refers to as "Rewrite Math" (2015). There are two characteristics of this mathematics that are homologous with evolution, particularly that related to Epigenetic Inheritance. First, there is the role of the term 'zero', which acts as an attractor, or reference point for the data set of the Rewrite System. The other aspect of the math is that in order to introduce a new datum into the set the System must evaluate its relationship to all of the other data in the set. That is essentially how epigenetic inheritance works, new data being evaluated within the egg or sperm during the process of meiosis by an unknown mechanism. One possibility is that because life correlates with Fibonacci sequences, there is an algorithm within the germ cells that uses that as a guide for whether a new term fits into the Fibonacci sequences of the DNA code or not.

DNA knots, homeobox genes and cell-cell signaling

Experimental evidence that DNA 'knots' are altered by changes in cell-cell signaling derives from data regarding injury-repair. We know that under stress conditions cells produce Reactive Oxygen Species that may cause gene mutations and duplications. Such DNA modifications are then acquired by the genome, hypothetically altering the topology of DNA. The most compelling evidence for this occurrence lies in the fact that Homeobox genes that determine the superstructure of the organism are arrayed along the chromosome in the same way that they appear in the organism (Mark et al., 1997).

Pre-adaptations, lipids immersed in water, and knots

The cellular approach to evolution is predicated on serial pre-adaptations or exaptations, begging the question as to 'where and when the knots related to life began?' In the scenario for lipids immersed in water as the fundament of life, lipids on earth arose from both thermal vents in the sea floor, or as passengers on the asteroids that delivered frozen water to earth (Deamer, 2017), the lipids originating from the Cosmos (Salama, 2008), as mentioned above.

Initially, the lipids aligned perpendicularly to the water-air interface because they are amphiphiles, or molecules with oppositely charged ends. Their orientation 'downward' was due to the force of gravity (Claassen and Spooner, 1996), their positively-charged hydrophobic ends facing upwards, their negatively-charged hydrophilic ends facing downward, immersed in the water phase. In this configuration the lipid molecules behave like molecular antennae, responding to electromagnetic forces such as Pulsars and photons of light (Fels, 2009). Once the lipids form micelles, or spheres with semi-permeable lipid membranes, the effect of the electromagnetic force acts to stimulate cytoplasmic streaming as a phenotypic flow of energy. Subsequently, when cholesterol appeared in the cell membrane, it thinned the cell membrane, further facilitating cytoplasmic streaming as locomotion. Fast-forward to the invertebrate Ciona intestinalis, a unicellular ancestor of vertebrates in which the stem cells for the heart develop in the tail, later migrating into the body to form the heart (Davidson et al., 2006). It is within the realm of possibility that the beating of the tail translates into the beating of the heart. This process may have exapted from the earlier evolution of cytoplasmic streaming since the heart evolved as a muscularization of the vasculature-like the heart of an earthworm, perhaps due to the homologous 'cytoplasmic streaming' of blood causing shear stress, which can generate gene mutations and duplications by producing Reactive Oxygen Species (Storr et al., 2013). The one-chambered heart of the earthworm could have given rise to the two-chambered heart of fish, the three-chambered heart of amphibians and reptiles, and ultimately the four-chambered heart of birds and mammals due to the effect of beta-adrenergic agents like adrenaline. For if you delete the beta-adrenergic receptor from developing mice, they are born with a two-chambered heart. The beta-adrenergic receptor 'duplicated' (Aris-Borsou et al., 2009), i.e., amplified during the water-land transition, likely promoting the evolution of the heart from two chambers in fish, to three chambers in amphibians and reptiles, to four chambers in birds and mammals.

The above provides a 'knot', beginning with micelles forming a boundary between the Explicate and Implicate Orders, all the way through to the evolution of the lung, heart and kidney. For example, the genetics of heart development is connected to that of the fingers, recognized as Timothy Syndrome, exemplifying the phenotypic homology between these structures based on cytoplasmic streaming.

The deep significance of the homology between knots and evolution

As alluded to above, the mechanism of cellular evolution is predicated on exaptations of sequentially earlier and earlier genetic traits, working backwards from present-day physiology to the unicellular state. Starting with lung evolution proved to be advantageous because gas exchange is a very ancient property of life, which allowed tracing of the process all the way back to the unicell, begging the question as to what the exaptive prototype for the cell itself was? The only such unity seems to be the Singularity of nature (Torday, 2019) that existed prior to the Big Bang. The rationale is that the unicell has the capacity to express all of the states of being- homeostasis, mitosis and meiosis- each of which is determined by the cytoskeleton as the 'sensor' for prevailing environmental conditions. But the insight that knots and evolution share common properties suggests that perhaps even before the homology between the cell and the Singularity, the shared topology of knots and cells may have formed the ultimate homology for the distinction between the inanimate and animate.

The metaphysics of knots

The homology between mathematical and cellular knots provides insight to the continuum from the inanimate to the animate. The phenomenon of lateralization (McGilchrist, 2009) extends that even deeper into our psyche by realizing that our right and left cerebral hemispheres act to perceive the left and right sides of our bodies, respectively, resulting in the 'unknotting' of the ambiguity from which we originated due to the negative entropy that Schrodinger described in What is Life? (1944).

By crossing over from left to right, or right to left in perceiving our surroundings, we reconcile that ambiguity and subjectivity of our senses gave rise to evolutionarily seeing the environment in what David Bohm referred to as the Implicate Order, the true order of things (1980).

However, when we look one another in the eyes, we transcend that 'trompe l'oeil'. We instinctively recognize this when a mother looks into her newborn's eyes, or love at first sight, or when we dance together, one person moving forwards, the other backwards. This is what happened to Narcissus when he saw his reflection in a pool of water? Or an artist creating a novel, a painting, or a piece of music? I have experienced insights from time to time while shaving (Hormones and Reality).....perhaps it was due to looking in the mirror while holding cold steel to my throat that gave clarity, like what I am expressing here?

Left-right brain resolves newtonian mechanics as quantum mechanics in biology—a hypothesis

Previously, it has been shown that the atom and the cell are homologues (Torday, 2018b), both being simultaneously deterministic and probabilistic. The atom's nucleus and the electron are in homeostatic balance, and the first three quantum numbers for the Pauli Exclusion Principle are deterministic, whereas the fourth principle is probabilistic, being based on time. Likewise, the first two First Principles of Physiology—negative entropy and chemiosmosis—are deterministic, whereas the third principle-homeostasis—is probabilistic. Moreover, physiology exhibits other Quantum Mechanical homologies such as non-locality, coherence and wave collapse when seen through the 'lens' of cell-cell communication as its basis. These insights are significant when it is finally realized that the life cycle is constituted by the unicellular state, transiting from zygote to zygote based on epigenetic inheritance (Skvortsova et al., 2018), the organism behaving as an 'agent' for the collection of data that modify the DNA readout of the egg or sperm. That perspective offers a much broader understanding of the interrelationship of the animate and inanimate, leveraged by Quantum Mechanics as the universal basis for everything in the Cosmos, bar none. Hypothetically, it is the differential deliberations by the left and right brain based on classic physics that leaves the mathematical 'remainder' as Quantum Mechanical. In so doing, such data can then be interpreted by the genome as 'meaningful' for inclusion or exclusion to the epigenome.

Discussion

Conventionally, consciousness is thought of as 'voices in our heads', or external to us (Clark and Chalmers, 1998). Those perspectives derive from our sense that we are material beings, when in fact we are mediators for the flow of energy within and between us (Torday, 2021a). As such, consciousness is imbedded in our physiology, which has been constructed from the Cosmos through Symbiogenesis. Exactly how we elicit such information from our physiology is not clear, though differences between us due to endocrine mechanisms has been exposited in a recently published book (Torday, 2022). Differences in receptor density on cells explains differences in behavior, which is underscored by the fact that the endocrine system is under epigenetic control (Zhang and Ho, 2011). It is through this insight that complicated physiology has been successfully reduced to the unicell as the origin (Torday and Rehan, 2012). And beyond that, the deconvolution of emergent events in physiologic evolution has led to a deep understanding of how and why Quantum Mechanics is the origin of life (Torday, 2021b), which had remained an enigma up until now.

References cited

Aris-Brosou, S., Chen, X., Perry S.F. and Moon, T.W. 2009. Timing of the functional diversification of alpha- and beta-adrenoceptors in fish and other vertebrates. Ann. N.Y. Acad. Sci. 1163: 343–347.

Berner, R.A. 1999. Atmospheric oxygen over Phanerozoic time. Proc. Natl. Acad. Sci. U. S. A. 96: 10955–10957.

Bloch, K. 1992. Sterol molecule: structure, biosynthesis, and function. Steroids 57: 378–383.

Bohm, D. 1980. Wholeness and the Implicate Order. Routledge, London.

Case, R.M., Eisner, D., Gurney, A., Jones, O., Muallem, S. and Verkhratsky, A. 2007. Evolution of calcium homeostasis: from birth of the first cell to an omnipresent signalling system. Cell Calcium 42: 345–350.

Claassen, D.E. and Spooner, B.S. 1996. Liposome formation in microgravity. Adv Space Res. 17: 151–160.

Clack, J. 2012. Gaining Ground. Indiana University Press, Bloomington.

Clark, A. and Chalmers, D.J. 1998. The extended mind. Analysis 58: 7–19.

Davidson, B., Shi, W., Beh, J., Christiaen, L. and Levine, M. 2006. FGF signaling delineates the cardiac progenitor field in the simple chordate, Ciona intestinalis. Genes Dev. 20: 2728–2738.

Deamer, D. 2017. The Role of Lipid Membranes in Life's Origin. Life (Basel) 7: 5.

Demayo, F., Minoo, P., Plopper, C.G., Schuger, L., Shannon, J. and Torday, J.S. 2002. Mesenchymal-epithelial interactions in lung development and repair: are modeling and remodeling the same process? Am. J. Physiol. Lung Cell Mol. Physiol. 283: L510–L517.

Fels, D. 2009. Cellular communication through light. PLoS One 4: e5086.

Gasque, G. 2015. A calcium-dependent mechanism of neuronal memory. PLoS Biol. 13: e1002182.

Gould, S.J. and Vrba, E.S. 1982. Exaptation—a missing term in the science of form. Paleontology 8: 4–15.

Hubble, E. 1929. A relation between distance and radial velocity among extra-galactic nebulae. Proc. Natl. Acad. Sci. U. S. A. 15: 168–173.

Koch, C. 2018. What is consciousness. Nature 557: S8–S12.

Levin, M. 2005. Left-right asymmetry in embryonic development: a comprehensive review. Mech. Dev. 122: 3–25.

Lodish, H., Berk, A., Kaiser, C.A., Krieger, M., Scott, M.P., Bretscher, A., Ploegh, H. and Matsudaira, P. 2008. Cell Signaling I: Signal Transduction and Short-Term Cellular Processes. W.H. Freeman and Company, New York.

Mark, M., Rijli, F.M. and Chambon, P. 1997. Homeobox genes in embryogenesis and pathogenesis. Pediatr. Res. 42: 421–429.

McGilchrist, L. 2009. The Master and His Emissary. Yale University Press, New Haven.

McNair, J.B., Dennis, L. and Kauffman, L. 2018. The Merion Matrix. World Scientific, Singapore.

Porges, S.W. 2007. The polyvagal perspective. Biol. Psychol. 74: 116–143.

Rowlands, P. 2015. The Foundations of Physical Law. World Scientific, Singapore.

Sagan, L. 1967. On the origin of mitosing cells. J. Theor. Biol. 14: 255–274.

Schrodinger, E. 1944. What is Life? Cambridge University Press, Cambridge.

Skvortsova, K., Iovino, N. and Bogdanović, O. 2018. Functions and mechanisms of epigenetic inheritance in animals. Nat. Rev. Mol. Cell Biol. 19: 774–790.

Salama, F. 2008. PAH's in Astronomy—a Review. Organic Matter in Space. Proceedings I.A.U. Symposium 251: 357–365.

Storr, S.J., Woolston, C.M., Zhang, Y. and Martin, S.G. 2013. Redox environment, free radical, and oxidative DNA damage. Antioxid. Redox Signal. 18: 2399–2408.

Torday, J.S., Torday, D.P., Gutnick, J., Qin, J. and Rehan, V. 2001. Biologic role of fetal lung fibroblast triglycerides as antioxidants. Pediatr. Res. 49: 843–849.

Torday, J.S. 2003. Parathyroid hormone-related protein is a gravisensor in lung and bone cell biology. Adv. Space Res. 32: 1569–1576.

Torday, J.S. and Rehan, V.K. 2007. The evolutionary continuum from lung development to homeostasis and repair. Am. J. Physiol. Lung Cell Mol. Physiol. 292: L608–L611.

Torday, J.S. and Rehan, V.K. 2012. Evolution, Cell-Cell Signaling and Complex Disease. Wiley, Hoboken.

Torday, J.S. and Miller, W.B. Jr. 2016. On the evolution of the mammalian brain. Front. Syst. Neurosci. 10: 31.

Torday, J.S. and Miller, W.B. Jr. 2017. The resolution of ambiguity as the basis for life: A cellular bridge between Western reductionism and Eastern holism. Prog. Biophys. Mol. Biol. 131: 288–297.

Torday, J.S. and Rehan, V.K. 2017. Evolution, the Logic of Biology. Wiley, Hoboken.

Torday, J.S. 2018a. From cholesterol to consciousness. Prog. Biophys. Mol. Biol. 132: 52–56.

Torday, J.S. 2018b. Quantum Mechanics predicts evolutionary biology. Prog. Biophys. Mol. Biol. 135: 11–15.

Torday, J.S. 2019. The Singularity of nature. Prog Biophys Mol Biol. 142: 23–31.

Torday, J.S. 2021a. Cellular evolution as the flow of energy. Prog. Biophys. Mol. Biol. 167, 147–151.

Torday, J.S. 2021b. Life is a mobius strip. Prog. Biophys. Mol. Biol. 167: 41–45.

Whitehead, A.N. 1979. Process and Reality. Free Press, Florence.

Zhang, X. and Ho, S.M. 2011. Epigenetics meets endocrinology. J. Mol. Endocrinol. 46: R11–R32.

Zheng, W., Wang, Z., Collins, J.E., Andrews, R.M., Stemple, D. and Gong, Z. 2011. Comparative transcriptome analyses indicate molecular homology of zebrafish swimbladder and mammalian lung. PLoS One 6: e24019.

Chapter 11
The Holism of Cosmology and Consciousness

"There are more things in heaven and Earth, Horatio, Than are dreamt of in your philosophy"

Wm. Shakespeare, "Hamlet"

Introduction

The greatest unsolved mystery of all is what consciousness is. It has been surmised that it is the homology or 'equivalent' of our physiology, behaving as our history (Torday and Miller, 2018a), providing an algorithm by which evolution can determine how and why to adapt to ever-changing environmental conditions. But of course, those ever-changing conditions are a consequence of an ever-expanding Cosmos (Penzias and Wilson, 1965), inferring a linkage between our consciousness and that of the Cosmos. But just how did this come about?

Complicated physiology was reduced iteratively to the unicell (Torday and Miller, 2016a) by tracing the cell-cell signaling mechanisms that dealt with emergent threats based on Bayesian statistics. Unlike Darwinian random mutations, the current approach is predicated on structural and functional changes within the context of pre-existing physiologic traits. For example, the lung evolved from the swim bladder of boney fish, founded not on function, which is now recognized as teleology (Roux, 2014), but on cellular molecular homologies that can be traced all the way back to the unicellular "First Principles of Physiology" (Torday and Rehan, 2009). By identifying key emergent changes over the course of evolution, and determining what pre-existing physiologic traits were exploited for adaptation (Gould and Vrba, 1982), the evolution of the lung was traced back to the semi-permeable cell membrane, particularly how and why lipids enhanced the evolution of the unicell. Conventionally, Conrad Bloch had reasoned after the fact that there must have been enough oxygen in the atmosphere to synthesize cholesterol, the latter requiring 11 atoms of oxygen to do so (Bloch, 1992). But that is reasoning after the fact, which is illogical. Conversely, there is a long history for the combined use of lipids and oxygen in the evolution of vertebrates (Torday and Rehan, 2004), beginning (actually ending) with the use of lipids to reduce the surface tension of the watery film covering the alveolar wall (Rooney, 1984). So, faced with the existential threat of rising oxygen in the atmosphere, lipids were yet again invoked, cholesterol being interposed within the phospholipid bilayer of the cell membrane, reducing the thickness of the barrier, increasing oxygenation and metabolism as a result, all in service to homeostasis, provided by Symbiogenesis, referencing Quantum Entanglement. This cascade is a consequence of the fractal formation due to the latter as local and non-local.

The co-opting of pre-existing traits in order to evolve is referred to as Symbiogenesis. Margulis borrowed this concept from the early 20th Century Botanist Mereschkowski (Mereschkowski, 1910). She implemented the concept of Symbiogenesis in order to explain how bacteria were assimilated to form mitochondria. Since then, numerous examples of substances in the environment posing existential threats being assimilated via Symbiogenesis have been demonstrated, acting to catalyze the evolution of multicellular organisms (Torday and Rehan, 2012).

That raises the question as to how and why Symbiogenesis arose in the first place as an emergent property? Based on the same Bayesian Theorem, energy for Quantum Entanglement of particles within the protocell was provided by gravitational force on the curved surface (Einstein, 1961), referencing Quantum Mechanics (QM) as the impetus for life. Thus, the cell offers the entire history of the organism from QM to the present ontogenetically and phylogenetically, offering options for adaptation to an ever-changing environment. According to the mathematician Garnet Ord (2013), the gamut from QM to Newtonian Mechanics represents Einstein's 'time dilation'. Thus, the cell is an organic proxy for the Relativity of spacetime.

Indeed, the cytoskeleton of the cell is a complete representation for all of the phases of life, past, present and future, controlling the Target of Rapamycin gene, which determines all of the stages of life—homeostasis (Torday, 2015a), meiosis and mitosis (Pennisi, 1998). Viewed from this perspective, the cell remains at equipoise relative to the Singularity that existed prior to the Big Bang (Torday, 2019a). The cell's purpose is to maintain that equipoise by running as fast as it can to remain as rest, like the Red Queen in "Alice in Wonderland". That view of life is a radical departure from Darwinian "Survival of the Fittest", which is an epiphenomenon. The former is diachronic, the latter is synchronic. Seen in this light, the purpose of life is to collect epigenetic marks (Torday and Miller, 2016b) by pursuing energy flows (Torday, 2021), not conventional material change due to random mutations.

The cell as the 'embodiment' of quantum mechanics

The cytoskeleton universally controls all of the states of the cell—homeostasis, meiosis and mitosis—rendering the status of the cell relative to the prevailing circumstances. The fact that the cell reprises its unicellular state over the course of its lifecycle is consistent with the mechanism of epigenetic inheritance, the organism behaving as the 'agent' for collecting 'marks' in the environment indicative of change (Torday and Miller, 2016b). Such marks are then assimilated by the organism, biochemically modifying stretches of DNA code through methylation or ubiquitination, for example (Zhang et al., 2021). Such modifications do not change the code itself, but change the 'readout' in the embryo, leading to phenotypic changes that allow the offspring to adapt.

The way in which the environment impacts the conceptus is via the cytoskeleton, which is monitored by the Target of Rapamycin gene (Jacinto et al., 2004), which is sensitive to cytoskeletal distortions. The mTOR gene regulates all aspects of the cell—metabolism, meiosis, mitosis—acting as a 'master gene'. The best example of that is that it affects lifespan in mice (Harrison et al., 2009).

The purpose of life is to monitor the environment for changes relevant to the organism

As a sensor for environmental change, this function refers all the way back to the differential between internal negative entropy (Schrodinger, 1944) and external positive entropy. It is the binary way of monitoring the environment that gives the cell the ability to register change in the environment effectively in a timely manner, prior to birth. The depth and width of that process was addressed above, linking the cell to the Cosmos, recognizing that consciousness is based on the Laws of Nature, forming a common denominator for all of existence, inanimate and animate alike (Torday, 2021). A classic example of such insight is the article by Edmund Husserl (1939), "The Origin of Geometry", which shows that geometry is not innate to us, it is acquired from the Cosmos. That is in contrast to consciousness, which is innate, based on our physiology, which is an algorithm for our 'history' (Torday, 2019b), having been formulated by the cell based on Symbiogenesis (Sagan, 1967), which is the acquisition of factors in the environment that have posed existential threats, being 'neutralized' by assimilating them, and compartmentalizing them, linked together through cell-cell communication, mediated by soluble growth factors as what we think of as physiology (Torday and Rehan, 2017).

On the origin of the enlarged human head/brain

In an earlier article (Torday, 2015b), it was hypothesized that physiological stress facilitated the transition from water to land by stimulating adrenalin production by the adrenal gland, hypoxia being the most potent physiologic agonist. That relieved the constraint on lung oxygenation by stimulating the production of lung surfactant, allowing the alveoli to further expand. As a 'side-effect', adrenalin also caused dissolution of fat cells, releasing free fatty acids into the circulation; free fatty acids are the most efficient source of metabolic energy, allowing for bipedalism since it takes more energy to do so than to crawl around on all-fours (Rodman et al., 1980). Bipedalism had a dramatic impact on human evolution, freeing the forelimbs for toolmaking and language, both oral and written. That combination of effects on the central nervous system created positive selection pressure for the progressive enlarging of the head and its contents, i.e., the brain (Hofman, 2014). Ultimately, the rate-limiting step was the capacity of the birth canal to accommodate the over-seized head at delivery, leading to relatively pre-term birth in homo sapiens. The net result was the birth of human newborns with only about 30% of adult brain capacity (Dunsworth, 2016), making the species ever-increasingly dependent upon the nuclear and extended family, and on society as a whole (Kaplan et al., 2009). That was facilitated by language, oral and written alike, but more so the latter as a way to perpetuate and inculcate the traditions of society. The problem with that is that our ambiguous origin gave rise to deceptive practices in order to cope with it (Trivers, 2014), and as a consequence, we periodically experience what Bateson et al. referred to as the 'double-bind' (Bateson et al. 1956), getting trapped in our deceits as a consequence of lying to ourselves and others.

Meditation as an end-run on deception

In considering the above origin of the enlarged human head/brain as a consequence of physiologic stress on metabolism, giving rise to endothermy may underlie the effect of meditation on hypometabolism (Flood, 1996).

Yogis have long been known to have the capacity to regulate their metabolism at will (Chaya et al., 2006). Formal study of this phenomenon has validated it scientifically. Functionally linking to ever-deeper principles of physiologic evolution through meditation and bio-feedback may prove to be of wider benefit in healing, both conventional and self-healing alike.

Life as meaning in the cosmos

The problem with the above is that there is seeming equanimity between the animate and inanimate. But that couldn't be further from the truth because it was only when lipids coalesced to form micelles, creating an inside and outside, forming the Explicate Order from the Implicate Order. It was the Explicate Order that provided the opportunity for life by distinguishing it from the inanimate Implicate Order (Bohm, 1980).

The Greek Protagaras thought that 'Of all things man is the measure' (Silvermintz, 2016). The position formulated above would have us understand that 'the cell is the measure'.

Discussion

The above-described integration of physiology as the cipher for consciousness is quite ingenious as a 'top-down/middle-out/bottom-up' way of ensuring the holism of life and non-life. Its novelty derives from the focus on the unicell as the source of that dynamism, acting as a mediator for factors passing through the semi-permeable membrane of the cell. The key to understanding the centrality of the cell for evolution is that it is composed of lipids that have molecular memory, or hysteresis. In order to be able to evolve effectively in response to the environment, the organism must rely upon such memory in order to be able to respond effectively to its ever-changing environment. DNA is conventionally thought of as the 'memory' of the cell, but in reality it appeared after the epigenetic property of the cell as the genetic determinant of memory (Torday, 2015c).

In the context of the above, it should be mentioned that both the cell [see Figure] (Torday, 2003) and the planets (Wurm and Teiser, 2021) are formed by the force of gravity. In the case of the cell, microgravity experimentally causes altered formation of micelles (Claassen and Spooner, 1996) in association with loss of cellular phenotypic identity (Torday, 2003; Purevdorj-Gage et al., 2006), reflecting a fundamental interrelationship between the force of gravity, the size of the Earth, cellular physiology and evolution. In the past, such interrelationships have been highlighted by Romer's 'greenhouse' effect (Romer, 1949) and Berner's data regarding the composition of atmospheric oxygen (Berner et al., 2007) and carbon dioxide (Berner, 1990). Suffice to say that it is how and why organisms have reacted to such events at the cellular-molecular level that have driven evolution (Torday and Rehan, 2012; Torday and Rehan, 2017), not Darwinian 'descent with modification' and 'natural selection', which are merely descriptive proxies for the former, which

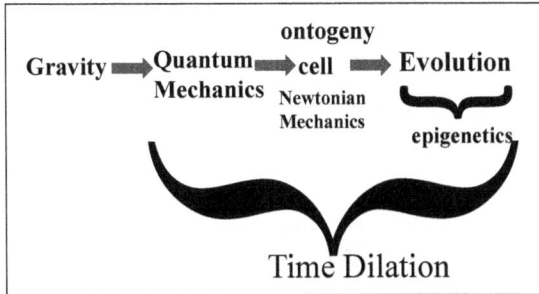

Figure. Role of Gravity in Evolution. From left to right, gravity manifests Quantum Mechanics as the basis for classical Newtonian Mechanics to maintain and sustain homeostasis. In the aggregate, that epigenetic serial process is described as ontogeny as the basis for Evolution.

offer falsifiable experimentation for the evolution of the glucocorticoid receptor (Bridgham et al., 2006), the thyroid gland (Nilsson and Fagman, 2017), and the lung (Torday, 2014) instead of associations and correlations, which do not show causation.

Recognition of the centrality of the cell has led to a number of novel insights into biology that have been represented by dogma up until now. Such otherwise conventional concepts as the life cycle (Torday, 2016a), phenotype (Torday and Miller, 2016b), homeostasis (Torday and Rehan, 2007), pleiotropy (Torday, 2018a), consciousness (Torday, 2018b), terminal addition (Torday and Miller, 2018b), niche construction (Torday, 2016b), heterochrony (Torday, 2016c) have all been redefined in light of the recalibration of the cell in biology. Consequently, our understanding of physiology has been simplified by shifting from description (Roux, 2014) to mechanism. Therefore, based on Ockam's Razor, the cellular approach to evolution is simpler, and is therefore superior to Darwin.

The above is on par with the realization that the Sun was the center of the Universe, which was a game changer.

Acknowledgements

John S. Torday has been supported by NIH Grant HL055268.

References cited

Bateson, G., Jackson, D.D., Haley, J. and Weakland, J. 1956. Toward a theory of schizophrenia. Behavioral Sci. 1: 251–264.

Berner, R.A. 1990. Atmospheric carbon dioxide levels over phanerozoic time. Science 249: 1382–1386.

Berner, R.A., Vandenbrooks, J.M. and Ward, P.D. 2007. Evolution. Oxygen and evolution. Science 316: 557–558.

Bloch, K. 1992. Sterol molecule: structure, biosynthesis, and function. Steroids 57: 378–383.

Bohm, D. 1980. Wholeness and the Implicate Order. Routledge, United Kingdom.

Bridgham, J.T., Carroll, S.M. and Thornton, J.W. 2006. Evolution of hormone-receptor complexity by molecular exploitation. Science 312: 97–101.

Chaya, M.S., Kurpad, A.V., Nagendra, H.R. and Nagarathna, R. 2006. The effect of long term combined yoga practice on the basal metabolic rate of healthy adults. BMC Complement. Altern. Med. 31: 28.

Dunsworth, H.M. 2016. Thank your intelligent mother for your big brain. Proc. Natl. Acad. Sci. USA 113: 6816–6818.

Einstein, A. 1961. Relativity. The Special and General Theory. Crown Publishing, New York.

Flood, G.D. 1996. An Introduction to Hinduism. Cambridge University Press, Cambridge.

Gould, S.J. and Vrba, E.S. 1982. Exaptation—A missing term in the science of form. Paleobiolgy 8: 4–15.

Harrison, D.E., Strong, R., Sharp, Z.D., Nelson, J.F., Astle, C.M., Flurkey, K., Nadon, N.L., Wilkinson, J.E., Frenkel, K., Carter, C.S., Pahor, M., Javors, M.A., Fernandez, E. and Miller, R.A. 2009. Rapamycin fed late in life extends lifespan in genetically heterogeneous mice. Nature 460: 392–395.

Hofman, M.A. 2014. Evolution of the human brain: when bigger is better. Front. Neuroanat. 8: 15.

Husserl, E. 1939. The origin of geometry. Rev. Int. Philos. 1: 157–179.

Jacinto, E., Loewith, R., Schmidt, A., Lin, S., Rüegg, M.A., Hall, A. and Hall, M.N. 2004. Mammalian TOR complex 2 controls the actin cytoskeleton and is rapamycin insensitive. Nat. Cell Biol. 6: 1122–1128.

Kaplan, H.S., Hooper, P.L. and Gurven, M. 2009. The evolutionary and ecological roots of human social organization. Philos. Trans. R. Soc. Lond. B Biol. Sci. 364: 3289–3299.

Mereschkowski, K. 1910. Theory of two types of plasms as the basis of symbiogenesis, a new study of the origin of organisms [part 1 of 4]]. Biologisches Centralblatt 30: 278–288.

Nilsson, M. and Fagman, H. 2017. Development of the thyroid gland. Development 144: 2123–2140.

Ord, G.N. 2013. The Physics of Reality. World Scientific, Hackensack.

Pennisi, E. 1998. Cell division gatekeepers identified. Science 279: 477–478.

Penzias, A.A. and Wilson, R.W. 1965. A measurement of excess antenna temperature at 4080 Mc/s. Astrophysical J. Letters 142: 419–421.

Purevdorj-Gage, B., Sheehan, K.B. and Hyman, L.E. 2006. Effects of low-shear modeled microgravity on cell function, gene expression, and phenotype in *Saccharomyces cerevisiae*. Appl. Environ. Microbiol. 72: 4569–4575.

Rodman, P.S. and McHenry, H.M. 1980. Bioenergetics and the origin of hominid bipedalism. Am. J. Phys. Anthropol. 52: 103–106.

Romer, A.S. 1949. The Vertebrate Story. University of Chicago Press, Chicago.

Rooney, S.A. 1984. Lung surfactant. Environ. Health Perspect. 55: 205–226.

Roux, E. 2014. The concept of function in modern physiology. J. Physiol. 592: 2245–2249.

Sagan, L. 1967. On the origin of mitosing cells. J. Theor. Biol. 14: 255–274.

Schrodinger, E. 1944. What is Life? Cambridge University Press, Cambridge.

Silvermintz, D. 2016. Protagoras. Bloomsbury Publishing, New York.

Torday, J.S. 2003. Parathyroid hormone-related protein is a gravisensor in lung and bone cell biology. Adv. Space Res. 32: 1569–1576.

Torday, J.S. and Rehan, V.K. 2007. The evolutionary continuum from lung development to homeostasis and repair. Am. J. Physiol. Lung Cell Mol. Physiol. 292: L608–L611.

Torday, J.S. and Rehan V.K. 2009. Lung evolution as a cipher for physiology. Physiol. Genomics 38: 1–6.

Torday, J.S. and Rehan, V.K. 2012. Evolutionary Biology, Cell-Cell Signaling, and Complex Disease. Wiley, Hoboken.

Torday, J.S. 2014. On the evolution of development. Trends Dev. Biol. 8: 17–37.

Torday, J.S. 2015a. Homeostasis as the Mechanism of Evolution. Biology (Basel) 4: 573–590.

Torday, J.S. 2015b. A central theory of biology. Med. Hypotheses 85: 49–57.

Torday, J.S. 2015c. The cell as the mechanistic basis for evolution. Wiley Interdiscip. Rev. Syst. Biol. Med. 7: 275–284.

Torday, J.S. 2016a. Life is simple-biologic complexity is an epiphenomenon. Biology (Basel) 5: 17.

Torday, J.S. 2016b. The cell as the first niche construction. Biology (Basel) 5: 19.

Torday, J.S. 2016c. Heterochrony as diachronically modified cell-cell interactions. Biology (Basel) 5: 4.

Torday, J.S. and Miller, W.B. 2016a. The unicellular state as a point source in a quantum biological system. Biology (Basel) 5: 25.

Torday, J.S. and Miller, W.B. 2016b. Phenotype as agent for epigenetic inheritance. Biology (Basel) 5: 30.

Torday, J.S. and Rehan, V.K. 2017. Evolution, the Logic of Biology. Wiley, Hoboken.

Torday, J.S. 2018a. Pleiotropy, the physiologic basis for biologic fields. Prog. Biophys. Mol. Biol. 136: 37–39.

Torday, J.S. 2018b. From cholesterol to consciousness. Prog. Biophys. Mol. Biol. 132: 52–56.

Torday, J.S. and Miller, W.B., Jr. 2018a. Unitary Physiology. Compr. Physiol. 8: 761–771.

Torday, J.S. and Miller, W.B., Jr. 2018b. Terminal addition in a cellular world. Prog. Biophys. Mol. Biol. 135: 1–10.

Torday, J.S. 2019a. The Singularity of nature. Prog. Biophys. Mol. Biol. 142: 23–31.

Torday, J. 2019. Evolution, the 'mechanism' of big history - the grande synthesis. J. Big History III: 17–24.

Torday, J.S. 2021. Cellular evolution as the flow of energy. Prog. Biophys. Mol. Biol. 167: 147–151.

Trivers, R. 2014. The Folly of Fools. Basic Books, New York.

Wurm, G. and Teiser, J. 2021. Understanding planet formation using microgravity experiments. Nat. Rev. Phys. 3: 405–421.

Zhang, Y., Sun, Z., Jia, J., Du, T., Zhang, N., Tang, Y., Fang, Y. and Fang, D. 2021. Overview of histone modification. Adv. Exp. Med. Biol. 1283: 1–16.

Chapter 12

Evolution, Gravity, and the Topology of Consciousness

Introduction

Consciousness is the aggregate of our evolutionary history, forged by ontogeny and phylogeny via cell-cell communication (Torday and Rehan, 2012). Cell-cell signaling provides the connection of physiology all the way back to the origin of life, and to the consciousness of the Cosmos as the way in which gravity initiated life, i.e., as lipid molecules submersed in the ocean, packing together, their negatively charged poles packing together to neutralize the surface tension of water.....that allowed for the quantum leap to micelles, bearing in mind that the lipid membrane is semipermeable, allowing for the flow of particles into and out of the protocell, the gravitational pull on the curved surface of the cell producing the energy needed to maintain the Quantum Entangled particles within it.

This way of thinking about our awareness of ourselves and our surroundings offers a solution to Chalmers' 'hard problem' (1995), for example, referencing existential threats from our evolutionary past. The following is a recounting of that process. But suffice to say that because our physiology is an encapsulation of our history, it provides a way of recalling events like our first witnessing of blood as 'seeing red'.

How and why evolution works has proven refractory to experimentation—how fish have evolved into Man, for example—because Darwinian evolution is based on random mutations, eliminating the possibility for testing such a causal mechanism, leaving only correlations and associations.

Ideally, there would be some cause-effect process that could be exploited in order to deconvolute evolution, but thus far none has been forthcoming. That is, up until now, having been introduced to Knot Theory (Kauffman, 2012), a subheading of Topology, which concerns itself with how geometric objects maintain themselves despite continuous deformations. Knot Theory is the study of mathematical knots, originating from literal physical knots. It has occurred to us that since knots tie things together, they are similar in kind to the cell-cell signaling mechanisms that 'tie' physiologic traits together during development, ultimately tying together the homeostatic forces that hold our physiology together (Cannon, 1963).

The formation of knots from a circle is homologous with the cell being 'tied into knots' to form our physiology. Bearing in mind that the math exists in the Cosmos, and has been dragged into our physiology as an 'artifact' (Husserl, 1939). The homology between the cell and the mathematical knot provides prima facie evidence that the conscious cell is the origin of both the Implicate and Explicate Orders (Bohm, 1980)—a conscious cell can conceive of a circle, but an unconscious circle cannot conceive of a cell.

Evolution is topologic because it began as a 'surface', formed by lipid micelles submersed in the ocean water that covered the primordial earth, generated by the force of gravity forming a boundary between the exterior Cosmos and the interior of micelles.

Lipids immersed in water spontaneously form micelles, lipid-bound spheres that are semi-permeable. The lipid molecules align at the air-water interface due to gravitational force, their negative ends pointing downward (hydrophilic), and once they generate enough negative charge locally, they break the surface tension due to the Van der Waals force for surface tension. That results in a quantum leap from molecules to the 'hol' of the micelle. The semipermeable nature of the micelle membrane allows for particles to move in and out, and the force of gravity on the curved surface of the micelle produces the energy for stabilizing the Quantum Entanglement of the particles, referencing the gravitational force of the Cosmos (= non-localization of Quantum Mechanics).

Such protocells comply with the First Principles of Physiology-negative entropy, chemiosmosis and homeostasis (Torday and Rehan, 2009a). Negative entropy runs counter to the positive entropy in the environment, constituting an ambiguous physical state of being (Torday and Miller, 2017). The processes of development and phylogeny inform the organism of environmental changes, the unicell twisting and turning in its purposeful efforts to conform with its environment in order to act as a phenotypic agent (Torday and Miller, 2016), collecting epigenetic 'marks' that keep the organism informed of environmental changes. Such data are then assimilated by the organism's germ cells, the egg and sperm, in order to inform the offspring to adapt.

In the case of mathematical knots, they are also generated from cell-like circles that 'twist and turn', held together by theoretical 'springs', and can be untied by releasing them, re-forming as circles. In the case of physiologic 'knots', they 'untie' themselves in response to injury, ultimately during aging or senescence (Torday and Rehan, 2012), returning to the unicellular state as a 'circle' over the course of the life cycle (Torday, 2016). So, like the test of a true mathematical knot being the ability to undo itself to form a circle, the 'test' of a biologic knot is the ability for it to revert back to its unicellular state as a zygote, and so on, and so forth iteratively, generation upon generation.

The homology between evolution and knots

As stated in the Introduction, knots are surfaces, which is how they are related homologically to evolution-lipids originating either from Pulsars or from thermal vents in the sea floor localizing to the air-water interface, where they formed micelles, prototypical cells that generated a surface between the Explicate and Implicate Orders described by David Bohm in his book "Wholeness and the Implicate Order" (1980), forming what Claude Bernard referred to as the 'internal milieu' (1974). This is the origin of unicellular organisms, which dominated the earth for the first 3.5 billion years. Due to competition between prokaryotes and eukaryotes, the latter began forming multicellular organisms through cell-cell communications mediated by growth factor-growth factor receptor signaling mechanisms. Such cellular pathways for growth and differentiation during embryologic development subsequently became the homeostatic basis for physiology, as alluded to above. In the process of doing so, the cells involved produce an extracellular matrix that physically stabilizes them (Kim et al., 2011), forming true 'knots' that literally tie our physiology together as organisms. During processes of chronic disease, such cellular communications break down, reverting back to earlier developmental and phylogenetic homeostatic stages of evolution, stabilized by scar tissue. During the aging process, loss of bioenergy similarly leads to systematic loss of homeostatic control and deterioration in cell-cell signaling (Torday and Rehan, 2012), revealing the knot-like characteristics of organismal structure and function.

Over the course of vertebrate evolution, organisms have coped with their ever-changing environments by endogenizing factors in their surroundings—heavy metals, gases, ions, bacteria—referred to as the Endosymbiosis Theory (Margulis, 1971). Once such factors were internalized, they were compartmentalized using intracellular membranes, forming what is commonly referred to as cell physiology. Organisms became iteratively more complicated based on cellular communications, producing the phylogenetic changes of speciation. But it should be borne in mind that organisms always return to the unicellular state over the course of their lifecycle (Torday, 2016). The capacity to form and unform structure and function, all the way back to a 'circle' is also the criterion for a mathematical knot, attesting to the homology of biological and mathematical knots.

Circles, knots, cells and memory

Memory is essential for the process of evolution—in order to remain in sync with an ever-changing environment, the organism must remember its past ways of adapting as the most efficient way of surviving, referred to as exaptations (Gould and Vrba, 1982). It is commonly held that DNA is our biologic form of memory, but the use of nucleic acids for this purpose is derivative of the lipid-based micelle, which has molecular memory in the form of hysteresis. The capacity to deform when warmed by the sun, and re-form in the cold of night is the origin of memory. Like this most basic functional property of life, knots also have memory, because as complicated as they may become, e.g., Trefoils, Borromeans, etc., they remain simple circles at their origin, witnessed by that being the formal validation that they are in fact knots. The same holds true for a multicellular organism, because it ultimately returns to its unicellular form over the course of its life cycle, demonstrating that it, too, is a knot. By returning to the unicellular state, the organism provides a way of integrating epigenetic 'marks' into its physiology over the course of its ontogeny, explaining the cellular basis for evolution.

And short of regressing all the way back to the unicellular state, tissues can default to earlier stages in their ontogeny and phylogeny due to loss of cell-cell signaling (Guex et al., 2020). This mechanism is due to the cellular networking for development and homeostasis, providing the option of defaulting to an earlier physiologic state in order to maintain itself, allowing the organism the opportunity to reproduce and transfer its genome to the next generation.

DNA knots, Homeobox genes and cell-cell signaling

Evidence that DNA 'knots' are altered by changes in cell-cell signaling is not forthcoming. Yet we know that under various stress conditions cells produce Radical Oxygen Species that may cause mutations and duplications. Such DNA modifications are then acquired by the genome, hypothetically altering the topology of DNA. The best evidence for this lies in the fact that Homeobox genes, which determine the structure of the organism are arrayed along the chromosome in the same way that they appear in the organism (Mark et al., 1997).

Pre-adaptations, lipids immersed in water, and knots

The cellular approach to evolution is predicated on serial pre-adaptations or exaptations, raising the question as to where and when the knots related to life began? In the scenario for the role of lipids in water as the genesis of life, lipids on earth arose from both thermal vents in the sea floor, or as hitch-hikers on the asteroids that delivered frozen water to earth, the lipids originating from Pulsars, as mentioned above.

Initially, the lipids aligned perpendicularly to the water-air interface because they are polarized amphiphiles, their orientation 'downward' apparently being due to gravity, their positively-charged ends facing upwards, being hydrophobic, their negatively-charged hydrophilic ends immersed in the water phase. Given that configuration, they behave like molecular antennae, responding to electromagnetic forces such as Pulsars and photons of light (Fels, 2009). When the lipids form micelles, which are spheres with semi-permeable membranes, the effect of the electromagnetic force acts to stimulate cytoplasmic streaming as a phenotypic flow of energy. Later on during evolution, when cholesterol appeared in the cell membrane, it caused thinning of the membrane, further facilitating cytoplasmic streaming as locomotion. Fast-forward to Ciona intestinalis, a unicellular vertebrate ancestor in which the stem cells for the heart develop in the tail, and then migrate into the body to form the heart (Davidson et al., 2006). It is within the realm of possibility to think that the beating of the tail translates into the beating of the heart. This process may have exaptively evolved from the earlier evolution of cytoplasmic streaming since the heart evolved as a muscularization of the vasculature-like the heart of an earthworm, perhaps due to the homologous 'cytoplasmic streaming' of blood acting to cause shear stress, known to generate gene mutations and duplications by generating Radical Oxygen Species (Storr et al., 2013). The one chambered heart of the earthworm could have given rise to the two-chambered hearts of fish, the three-chambered hearts of amphibians and reptiles, and ultimately to the four-chambered hearts of birds and mammals due to the effect of beta-adrenergic agents like adrenaline. For if you delete the beta-adrenergic receptor from developing mice, they are born with a two-chambered heart. The beta-adrenergic receptor 'duplicated' (Aris-Borsou et al., 2009), i.e., amplified during the water-land transition, likely promoting the evolution of the heart from two chambers in fish, to three chambers in amphibians and reptiles, to four chambers in birds and mammals.

The above provides a 'knot' that begins with lipids forming a boundary between the Explicate and Implicate Orders, all the way through to the evolution of the lung, heart and kidney. The genetics of heart development, for example, is connected to that of the fingers, recognized as Timothy Syndrome, exemplifying the phenotypic homology between these structures based on cytoplasmic streaming.

The deep significance of the homology between knots and evolution

As expressed above, the mechanism of cellular evolution is predicated on exaptations from sequentially earlier and earlier genetic traits, working backwards from present-day physiology to the unicellular state. Starting with lung evolution proved to be advantageous because gas exchange is a very ancient property of life that allowed tracing of the process all the way back to the unicell. But that begs the question as to what the exaptive prototype for the cell itself was? The only such unity seems to be the putative Singularity of nature (Torday, 2019) that existed prior to the Big Bang. The rationale is that the unicell has the capacity to express all of the states of being—homeostasis, mitosis and meiosis—all three being determined by the cytoskeleton as the 'sensor' for prevailing environmental conditions. But the insight that knots and evolution share common properties suggests that perhaps even before the homology between the cell and the Singularity, the shared topology of knots and cells may have formed the ultimate homology for the inanimate and animate.

The metaphysics of knots

The homology of physical and cellular knots provides insight to the continuum from the inanimate to the animate. The phenomenon of lateralization extends that even deeper into our psyche by realizing that our right and left cerebral hemispheres act to perceive the left and right sides of our bodies, respectively, perhaps 'unknotting' the ambiguity from which we originated due to the negative entropy that Schrodinger described in What is Life? (1944).

By crossing over from left to right, or right to left in perceiving our surroundings, we are able to reconcile that ambiguity and the subjectivity of our senses it gave rise to evolutionarily, seeing the environment in what David Bohm referred to as the Implicate Order, the true order of things.

However, when we look one another in the eyes, we transcend that 'trompe l'oeil'. We instinctively recognize this when a mother looks into her newborn's eyes, or love at first sight, or when we dance together. Perhaps this is what happened to Narcissus when he saw his reflection in a pool of water? Or when an artist creates a novel or a painting or a piece of music? I have experienced insights from time to time while shaving..... perhaps it was due to looking in the mirror while holding cold steel to my throat that gave clarity, like what I am expressing here?

Left-right brain as the resolution of quantum mechanics in biology—a hypothesis

It has previously been shown that the atom and the cell are homologues (Torday, 2018), both of which are simultaneously deterministic and probabilistic. In the case of the atom, the nucleus and the electron are in homeostatic balance. Furthermore, the first three quantum numbers for the Pauli Exclusion Principle are deterministic, whereas the fourth principle is probabilistic, based on time. Similarly, the first two First Principles of Physiology-negative entropy and chemiosmosis—are deterministic, whereas the third principle-homeostasis—is probabilistic. Beyond such cellular characteristics, physiology has other Quantum Mechanical homologies such as non-locality, coherence and wave collapse when seen through the 'lens' of cell-cell communication as its basis. These insights are particularly significant when it is finally realized that the life cycle is constituted by the unicellular state, transiting from zygote to zygote based on epigenetic inheritance (Torday and Miller, 2016), the organism behaving as an 'agent' for the collection of data that modify the DNA readout of the egg or sperm. That perspective offers a much broader understanding of the interrelationship of the animate and inanimate, leveraged by Quantum Mechanics as the universal basis for everything in the Cosmos, bar none. Hypothetically, it is the differential deliberations by the left and right brain based on classic physics that leaves the mathematical 'remainder' as Quantum Mechanical. In so doing, such data can then be interpreted by the genome as 'meaningful' for inclusion or exclusion to the epigenome.

Discussion

Random gene mutations as the basis for Darwinian evolution disconnect it from development and phylogeny. Conversely, the reverse-engineering of cell-cell communications has offered fundamental insights to the 'how and why' of the ontology and epistemology of evolution. The current delineation of the interrelationship between cell-cell communication and Knot Theory further expands on such fundamental causal mechanisms, providing support for the previously expressed concept of "The Singularity of Nature" (Torday, 2019). It is only through such cross-cutting, transcendent diachronies that we will ultimately be able to form an algorithmic perspective for a holistic understanding of life in the Cosmos.

Up until now it has only been the artists who have been able to express the holism of our existence. For example, Frost said that 'life is the which can mix oil and water', or in his poem 'The Secret Sits'.... in the middle, and we dance around wondering why. Similarly, in the poem "Little Gidding", T.S. Eliot assures us that -

"We shall not cease from exploration And the end of all our exploring Will be to arrive where we started And know the place for the first time."

In art, Magritte's painting "Not to be Reproduced" [see Chapter 9, Figure 4] depicting a man seeing himself in a mirror from behind, expresses the sense of the observer and the observed, while Henry Moore confronts us with the question as to whether the negative space in his sculptures has equal or greater value than the granite or bronze? With the merging of evolution and topology there is the opportunity to solve these enigmas scientifically and mathematically at long last.

References cited

Aris-Brosou, S., Chen, X., Perry, S.F. and Moon, T.W. 2009. Timing of the functional diversification of alpha- and beta-adrenoceptors in fish and other vertebrates. Ann. N.Y. Acad. Sci. 1163: 343–347.

Bernard, C. 1974. Lectures on the Phenomena of Life Common to Animals and Plants. Charles C. Thomas, Springfield.

Bohm, D. 1980. Wholeness and the Implicate Order. Routledge, New York.

Cannon, W.B. 1963. The Wisdom of the Body. W.W. Norton, New York.

Chalmers, D. 1995. Facing up to the problem of consciousness. J. Consciousness Studies 2: 200–219.

Davidson, B., Shi, W., Beh, J., Christiaen, L. and Levine, M. 2006. FGF signaling delineates the cardiac progenitor field in the simple chordate, Ciona intestinalis. Genes Dev. 20(19): 2728–38.

Deamer, D. 2017. The Role of Lipid Membranes in Life's Origin. Life (Basel) 7(1): 5.

Fels, D. 2009. Cellular communication through light. PLoS One 4(4): e5086.

Gould, S.J. and Vrba, E.S. 1982. Exaptation—a missing term in the science of form. Paleobiology 8: 4–15.

Guex, J., Torday, J.S. and Miller, W.B. Jr. 2020. Morphogenesis, Environmental Stress and Reverse Evolution. Springer, Basel.

Husserl, E. 1939. The origin of geometry. Rev. Int. Philos. 1: 157–179.

Kauffman, L. 2012. Knots and Physics. World Scientific Publishing Company, Singapore.

Kim, S.H., Turnbull, J. and Guimond, S. 2011. Extracellular matrix and cell signaling: the dynamic cooperation of integrin, proteoglycan and growth factor receptor. J. Endocrinol. 209: 139–151.

Margulis, L. 1971. Symbiosis and evolution. Sci. Am. 225: 48–57.

Mark, M., Rijli, F.M. and Chambon, P. 1997. Homeobox genes in embryogenesis and pathogenesis. Pediatr. Res. 42: 421–429.

Storr, S.J., Woolston, C.M., Zhang, Y. and Martin, S.G. 2013. Redox environment, free radical, and oxidative DNA damage. Antioxid. Redox Signal. 18: 2399–2408.

Torday, J.S. and Rehan, V.K. 2009a. Lung evolution as a cipher for physiology. Physiol Genomics. 38: 1–6.

Torday, J.S. and Rehan, V.K. 2009b. The evolution of cell communication: the road not taken. Cell Commun. Insights. 9(2): 17–25.

Torday, J.S. and Rehan, V.K. 2012. Evolutionary Biology, Cell-Cell Communication and Complex Disease. Wiley, Hoboken.

Torday, J.S. 2016. The Cell as the First Niche Construction. Biology (Basel) 5: 19.

Torday, J.S. and Miller, W.B. 2016. Phenotype as agent for epigenetic inheritance. Biology (Basel) 5: 30.

Torday, J.S. and Miller, W.B. Jr. 2017. The resolution of ambiguity as the basis for life: A cellular bridge between Western reductionism and Eastern holism. Prog. Biophys. Mol. Biol. 131: 288–297.

Torday, J.S. 2018. Quantum Mechanics predicts evolutionary biology. Prog. Biophys. Mol. Biol. 135: 11–15.

Torday, J.S. 2019. The Singularity of nature. Prog. Biophys. Mol. Biol. 142: 23–31.

Chapter 13
Cellular Evolution and the Flow of Energy

Introduction

Einstein dreamt that he was chasing a beam of light as a 16 year old (Isaacson, 2008). That inspiration was the core of his Wonder Year of 1905, composed of three seminal papers—Brownian Movement, the Photoelectric Effect, and Special Relativity, respectively. That was a lens through which he viewed the Cosmos being composed of energy, and that nothing could travel faster than the speed of light. The following is the reduction of physiologic evolution to the flow of energy, formed from Einstein's insights. When evolution is seen as energy instead of matter it forms a solution to the 'forest and trees' problem, complying with Occam's Razor, that the simplest answer is the correct one. That simplicity is exemplified by the way that Quantum Entanglement acts to mimic the role of Symbiogenesis, maintaining homeostasis, forming the link between the physical and the biological, the local/non-local of entanglement providing the fractal infrastructure for all things Cosmos.

Whereas Darwinian evolution is predicated on material competition between organisms, cellular epigenetic evolution focuses on the transfer of data from one stage of life to another mediated cell-cell communication—developmental, phylogenetic—as injury-repair, obeying the First Principles of Physiology (Torday and Rehan, 2009). This evolutionary mechanism is founded on ontogeny as the only known mechanism for the formation of biologic structure and function. And ontogeny is the mechanistic basis for phylogeny as speciation (Torday and Rehan, 2012, 2017). And since both are based on cell-cell communication mediated by soluble growth factors and their receptors, ontogeny and phylogeny can be merged together as evolution, acting as one unified process [see Figure] (Torday and Rehan, 2012, 2017). In this light, the material aspects of the organism can be seen as 'means', not 'ends', questioning what

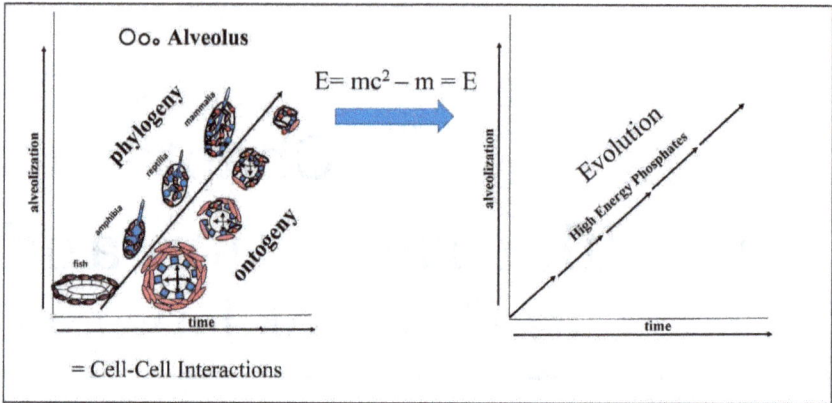

Figure. Ontogeny, Phylogeny and Evolution of the Lung. On the left-hand side are the phylogeny and ontogeny of the lung, emphasizing alveolarization (abscissa) as a function of cell-cell interactions (ordinate). On the right-hand side is the evolution of the lung as alveolarization (abscissa) versus time (ordinate) as a function of high-energy phosphate signaling, emphasizing the latter instead of the conventional 'material change' over the course of development.

the 'ends' of evolution are ? Once material aspects of life are eliminated, the flow of energy remains, both within and between generations. What are the ontologic and epistemologic significance of evolution when seen as energy flows?

The Big Bang as the flow of all things

The earth is about 4 billion years old, having formed from the Big Bang, which occurred about 13.8 billion years ago. The evidence that such a cataclysm occurred is based on the Redshift, the infrared radiation detected by Penzias and Wilson (1965). That was a huge breakthrough in our understanding of how and why the Cosmos formed from an initial source, the Singularity, before which Astronomy consisted of descriptions of various celestial phenomena without a central theory. In contrast to that, nowadays Lee Smolin has exploiting Darwinian evolution in order to explain the formation of Black Holes and stars. Significantly, the existence of an 'origin' for the Cosmos infers that there was a vectoral 'beginning', not unlike the origin of life as a unicell (Torday and Rehan, 2012), the submersion of lipids in water spontaneously generating micelles, providing a beginning for life. That transition of lipid molecules to micelles was quantum.

A priori, the Cosmos began as the explosive disruption of the Singularity, releasing energy. Homeostasis is the 'recoil' of that event, every action having an equal and opposite reaction ('Newton' Third Law of

Motion). Matter was the product of that homeostatic counterforce, energy being the primary state (Whithead, 1929). Therefore, in both principle and practice, it would behoove us to consider evolutionary change as energy instead of matter.

Complexity is an artifact of material evolution

Cell-cell communications mediate the formation of complicated phenotypes, mediated by high energy phosphates as 'second messengers'. Such messengers ultimately affect the read-out of DNA within the nucleus, translating it into RNA, and then to protein, which is considered the 'central dogma' of biology, the proteins affecting both the growth and differentiation of the affected cell. The targeted cell then produces soluble growth factors that affect neighboring cells, in turn causing either their growth or differentiation, etc. We focus on the material stages of this cascade of developmental forms, ignoring the underlying transfer of energy from cell to cell, considering it to be a means, and not an end. However, as alluded to above, merging ontogeny and phylogeny into one unified process eliminates the perceived spatial and temporal differences between the two based on descriptive biology. Now all that is left are the energetic changes from potential to kinetic, the forms being 'nodes' for the energetic 'pathways'. This perspective raises the question as to its origin. Having hypothesized that the unicell formed with reference to the Singularity (Torday, 2019), the physiologic energy vector appears to have been the consequence of the Big Bang.

Symbiogenesis, the means by which life has evolved

Unicellular life has existed on earth for 3.5 billion years. Multicellular organisms appeared approximately 500 million years ago. It has been hypothesized that multicellular organisms evolved in response to bacteria affecting pseudo-multicellular forms based on Biofilm and Quorum Sensing, which are ad hoc pseudostructures for multicellularity that formed to cope with the environment. In response to the existential threat to eukaryotic unicellular organisms, the latter evolved cell-cell communications mediated by soluble growth factors and their receptors, giving rise to multicellular organisms.

The prevailing theory of cellular evolution is Symbiogenesis, or the internalization of factors in the environment that presented as existential threats. Primordial cells spontaneously formed from micelles with semipermeable cell membranes that facilitated the acquisition of substances from the environment. Internal membranes compartmentalized such substances and organized them biochemically as the basis for physiology (Torday and Rehan, 2012, 2017).

The endogenized substances in the Cosmos are ordered based on the way that stars produce the Elements during the process of stellar nucleosynthesis, beginning on the far left of the Periodic Table with Hydrogen, progressing from left to right 'periodically' as heavier and heavier substances, having progressively more protons in their nuclei. In this way, the Elements form the hierarchical basis for the Laws of Nature, to which both the Cosmos and living organisms comply. Consequently, our consciousness is derived from the 'Consciousness' of the Cosmos (Torday, 2020).

'Phenotype as agent' addresses the significance of energy flow

Epigenetic inheritance is characterized by the capacity of the organism to identify specific changes in the environment, assimilate them biochemically in the egg and sperm of the ovaries and testes, respectively, and transfer such information to the offspring. As a result, the organism essentially has foreknowledge of environmental changes, and is able to modify its phenotype accordingly. The mechanism by which the embryo incorporates such epigenetic data obtained from the environment is dependent on cell-cell communications mediated by high energy phosphate 'second messengers' constrained by homeostasis. Therefore, the phenotype is behaving as an 'agent' for the faithful transfer of energy from one generation to the next in order to adapt to the ever-changing environment. Seen this way, phenotypic change is a means to an end, not an end in itself as Darwinian evolution would lead us to believe.

Of televisions, phenotypes and wiring

We have made systematic errors in judgement based on our Explicately evolved senses in the past, like the Earth being the center of the Solar System, the Earth being flat, or spontaneous generation. So think about your TV set, with images formed by electrons interacting with the rare earth coating on the inside of the screen. And then there are the wires in the back of the TV set that conduct the flow of electrons. It is there that the origin and principles involved emanate from. It is like the images on the screen are phenotypes, the electron flow being the basis for such forms.

Coherence to what?

In "Wholeness and the Implicate Order", David Bohm addresses how we cohere, but he does not tell us what we cohere to. Based on the precept that the Cosmos is expanding, by following the energetic vector formed by the Big Bang coherence would be to that vector. Deviating from the vector can occur, but it causes stress, which the organism can accommodate through auto-engineering, but there are limits. If the stress is too great, the organism will not be able to maintain its coherence, becoming extinct as a consequence.

Implications for focusing on evolution as energy flow

William of Ockham proposed that the simplest answer was correct. Conversely, Darwinian evolution gives rise to a complicated perspective for life—"It is interesting to contemplate a tangled bank, clothed with many plants of many kinds, with birds singing on the bushes, with various insects flitting about, and with worms crawling through the damp earth, and to reflect that these elaborately constructed forms, so different from each other, and dependent upon each other in so complex a manner, have all been produced by laws acting around us. These laws, taken in the largest sense, being Growth with reproduction; Inheritance which is almost implied by reproduction; Variability from the indirect and direct action of the conditions of life, and from use and disuse; a Ratio of Increase so high as to lead to a Struggle for Life, and as a consequence to Natural Selection, entailing Divergence of Character and the Extinction of less improved forms. Thus, from the war of nature, from famine and death, the most exalted object which we are capable of conceiving, namely, the production of the higher animals, directly follows. There is grandeur in this view of life, with its several powers, having been originally breathed by the Creator into a few forms or into one; and that, whilst this planet has gone circling on according to the fixed law of gravity, from so simple a beginning endless forms most beautiful and most wonderful have been, and are being evolved."

However, when seen as energy instead of matter, as a 'forest and trees' problem, evolution is much simpler. In fact, when the criterion of evolution as 'serial pre-adaptations', or exaptations is applied, there are many stops along the way where the connections prove impossible based on material appearance. Take the example of the homology between the swim bladder of a fish and the lung of a mammal. The analogy is between the gills and the lung, both mediating gas exchange, as has been the case for decades. But with the advent of functional genomics, it has become clear that the homology (or being of the same origin), is that between the fish swim bladder and mammalian lung. Similarly, there had been a glitch in pursuit of the evolution of the Central Nervous System, until Nicholas Holland pointed out that worms have their nervous systems in their skin (Holland, 2003), providing a way to bridge the gap between invertebrates and vertebrates. Moreover, there are deep cellular-molecular homologies between the skin and brain that arise from the mechanism for the formation of the stratum corneum as a barrier against bacterial invasion from the outside and loss of fluid and electrolytes from the vasculature. The mechanisms for depositing lipids as a barrier and host defense peptides in the skin is homologous with mechanism of secreting lung surfactant in the alveoli, and the myelination of neurons by Schwann Cells. Such molecular homologies attest, for example, to the causal relationships for patients with neurodegenerative diseases like Tay Sachs, Niemann-Pick and Schizophrenia, all of which have atopic dermatitis, or Atopy, a skin lesion. Similarly, patients with Asthma also have Atopy. The genetic link between these seemingly disparate traits is Parathyroid Hormone-related Protein (PTHrP), which links all of these phenotypes. Importantly, the PTHrP Receptor 'duplicated' during the water-land transition, amplifying that signaling pathway in a variety of tissues and organs, making terrestrial life feasible, among them being the lung, kidney, skin and brain.

Exemplary for understanding such interrelationships is the controversy over Darwinian gradualism and Eldredge and Gould's 'punctuated equilibrium' (Gould and Eldredge, 1993). Similar to Bohr's explanation for the duality of light as both a particle and wave, it is a function of how you measure it. From a molecular perspective, the same evolutionary trait initially occurs as a microscopic or sub-microscopic change, but over time it becomes visible, hence the difference in perspectives between Darwin, on the one hand, and Eldredge and Gould on the other.

Albert Szent-Gyorgyi, was a founder of modern biochemistry, who said that life is an interposition between two energy levels of an electron: the ground state and the excited state. There is evidence that light can stimulate cell division in paramecia, for example. And the observation that retinal cells can detect single photons lends credence to the role of Quantum Mechanics in biology.

Insights gleaned from evolution as energy flow

The overarching hypothesis is that the disruption of the Singularity by the Big Bang gave rise to a vectoral flow of energy that generated the Cosmos, and subsequently gave rise to life. That perspective naturally lends itself to the continuum from the physical to the organic, as has previously been hypothesized, the atom and the cell both act as point sources, being homologues, exhibiting both deterministic and probabilistic characteristics. This reductionist perspective lends itself to reconsidering the very nature of evolution, particularly given the re-emergence of epigenetic inheritance based on the primacy of the unicellular state—epigenetic marks are assimilated by the egg and sperm, and subsequently incorporated into the offspring. Consequently, we now recognize that we never leave the unicellular state, instead delegating the offspring as an 'agent' for collecting epigenetic data from the environment. Seen this way, evolutionary adaptation optimizes the detection of environmental perturbations, emphasizing changes in energy flow as the characteristic to be monitored. The phenotypic changes that result are meant to rectify such changes in energetic flow.

Darwinian evolution has heavily influenced contemporary thought, from sociology to psychology, literature, and astronomy. Yet we continue to dither in search of ways to formulate a 'Theory of Everything' (Weinberg, 2011) as the only organism that is destroying ourselves and the planet. It is proposed that that's because we have focused on the material aspect of life, when in fact it is the energy side of $E = mc^2$ that holds the answer.

If, for example, literature focused on energy, perhaps it would be more consistent with the vector of the Big Bang and the 'universality' it strives for. Similarly, instead of focusing on 'supply and demand', economics could be centered on the amount of Free Energy in the system, consistent with its relationship to physiology. And history would be taught based on the Free Energy available to society, rather than on personalities and philosophies, which would be closer to what David Christiansen's history is attempting (2018).

Discussion

The most compelling argument against Darwinian evolution is that it is illogical, reasoning after the fact based on descriptive biology. In lieu of any other means of understanding evolution, Darwinism has prevailed up until now. However, with the resurgence of epigenetic inheritance, particularly as it applies to evolution (Torday and Rehan, 2012, 2017), an alternative way of considering evolution from its unicellular origins is now available.

Historically, the major advances in perceiving our environment have countered common sense perceptions thanks to empiricism — Copernican Heliocentrism as the Sun being the center of the solar system, the Earth being round, life as spontaneous generation. Similarly, the present hypothesis that evolution is the flow of energy, not Darwin's 'tangled bank', is likewise founded on experimental evidence (Torday and Rehan, 2012).

The breakthrough in our understanding of embryonic development was the discovery of soluble growth factors as paracrine signals for structure and function, mediated by second messengers as high energy phosphate compounds. Once that mechanism was extended to phylogeny as the long-term history of speciation, it was a solution for understanding evolution from its origins instead of its consequences (Torday and Rehan, 2012) — Seeing evolution in the forward direction, from the unicell forward provided a causal relationship emanating from its physical origins in the Big Bang (Torday, 2019). Seen as a continuous process instead of as haphazard random mutations, it became clear that we should focus on evolution as serial energy exchanges, beginning with the Big Bang.

Contemporary biology and physics are at a critical phase, unable to reduce their problems to practice using their reductionist principles. In the case of biology, biomedical research is in a crisis, unable to bridge the gap between the gene and the phenotype. And as for physics, Quantum Mechanics is unable to explain everyday realities in the way that Newtonian physics has been able to. It appears that we have reduced these complex problems to absurdity, or reductio ad absurdum. The problem emanates from our after the fact rationalization of our own beginnings and evolution. Up until now, all we had was a compendium of organisms, beginning with Linnaeus's Binomial Nomenclature, that has now been reduced to genes. Yet genes do not form structures and functions, cells do. This error is due to a systematic error on the part of the evolutionists, deciding to side with geneticists in order to advance their knowledge, skipping over cell biology as the 'holism', to this day. As for the physicists, they advanced our knowledge mathematically, but they now realize that there is a seeming disconnect between Quantum Mechanics and day-to-day reality (Fauchiger and Renner, 2018).

This is a plea for continued scientific inquiry in an era when scientists are defaulting to spiritualism as the only seeming alternative. The Galileo Commission, for example, and the Non-duality of Science, or SAND group (www.scienceandnonduality.com), opt for belief instead of rational scientific method. This failure to pursue scientific quests based on hypothesis testing Popperian falsafiability is due to a lack of perspective and imagination.

References cited

Christiansen, D. 2018. Origin Story: A Big History of Everything. Little, Brown and Company, Boston.

Frauchiger, D. and Renner, R. 2018. Quantum theory cannot consistently describe the use of itself. Nat. Commun. 9: 3711.

Gould, S.J. and Eldredge, N. 1993. Punctuated equilibrium comes of age. Nature 366: 223–227.

Holland, N.D. 2003. Early central nervous system evolution: an era of skin brains? Nat. Rev. Neurosci. 4: 617–627.

Isaacson, W. 2008. Einstein: His Life and Universe. Simon and Schuster, New York.

Penzias, A.A. and Wilson, R.W. 1965. A measurement of excess antenna temperature at 4080 Mc/s. Astrophys. J. Lett. 142: 419–421.

Torday, J.S. and Rehan, V.K. 2009. Lung evolution as a cipher for physiology. Physiol. Genom. 38: 1–6.

Torday, J.S. and Rehan, V.K. 2012. Evolutionary Biology, Cell-Cell Communication and Complex Disease. Wiley, Hoboken.

Torday, J.S. and Rehan, V.K. 2017. Evolution, the Logic of Biology. Wiley, Hoboken.

Torday, J.S. 2019. The Singularity of nature. Prog. Biophys. Mol. Biol. 142: 23–31.

Torday, J.S. 2020. Consciousness, redux. Med. Hypotheses 140: 109674.

Weinberg, S. 2011. Dreams of a Final Theory: the Scientist's Search for the Ultimate Laws of Nature. Knopf Doubleday Publishing Group, New York.

Whitehead, A.N. 1929. Process and Reality. Macmillan, New York.

Chapter 14
Fractal Properties of Physiology
How and Why

Introduction

What follows is a rationale for this book using the concept of fractals as the holistic, integrating principle. Starting with Quantum Entanglement as the homolog for Symbiogenesis as the way in which the inanimate and animate relate to one another provides the local/non-local premise for consciousness that is at the center of the book. Everything literally emanates from there recursively and redundantly.

Local vs non-local consciousness

It has been hypothesized that physiology is actually the aggregate of our consciousness (Torday, 2020), largely founded on the effect of gravity on cellular identity (Torday, 2003). When consciousness is lost in a Near Death Experience, an Out of Body Experience, a Maslow Peak Experience, meditation, or the Runner's High, the local effect of gravity that confers physiology, or ego, is diminished or lost, causing the organism to default to the non-local gravity of the Cosmos. Similarly, as we age or senesce, there is a loss of cell-cell signaling due to decline in growth factor receptors (Balasubramanian and Longo, 2016), leading to a gradual loss of the local effect of gravity, defaulting to the non-local gravity of the Cosmos.

Although hard evidence for the homeostatic balance between the local and non-local gravitational effects is not forthcoming, there is circumstantial evidence like the recapitulation of brain evolution among patients undergoing general anesthesia. Or the Libet Experiment, in which subjects undergoing electroencephalogram monitoring were administered an electrical shock. It was found that the stimulation of brain activity anticipated the physical reaction by 300 milliseconds, inferring that the mind knows how to react before the body does.

Even the 'Double-slit' experiment, shining a light on a template with two slits in it, generating two patterns when seen by the observer. But when we look away, the light beam is one continuous path. It is proposed that the image of particles displayed on the wall is manifests the digital of the cell membrane, whereas the continuum of a light wave manifests our physiology, 'remembering' our history.

"Body memory" and consciousness?

Currently, cellular memory speculates that body cells contain information about our psyche's likes and dislikes, and our history unrelated to DNA or the brain. However, the relatively recent discovery that the endocrine system is under epigenetic control (Zhang and Ho, 2011) is a game changer, inferring that the effects of the hormones that control our emotions can be directly inherited from the environment.

Epigenetic Inheritance occurs when some factor in our environment is interpreted as an existential threat, and is assimilated in the egg or sperm as 'epigenetic marks', which may alter the development of the offspring such that it can adapt, consistent with the hypothesis that the organism is an 'agent' for detecting changes in its environment. This mechanism of inheritance is also consistent with Terminal Addition, the adding on of newly evolved traits at the end of a series of evolved traits, given that such traits are determined by cell-cell signaling. To add a new trait in the middle or at the beginning of such series of evolved traits would be highly inefficient, given that the mechanism entails upstream effects of 'second messengers' that affect growth and differentiation.

In "Molecules of Emotion", Candace Pert expressed the idea that "the mind is not just in the brain, but also exists throughout the body." She proffered that "The mind and body communicate with each other through chemicals known as peptides". "These peptides are found in the brain as well as in the stomach, muscles….in fact, all of our major organs and tissues". Pert believed that memory could be accessed anywhere in the peptide/hormone/receptor network. For instance, a memory associated with food might be linked to the pancreas or liver, and such associations could be transplanted from one person to another epigenetically.

Pert's notions have not found favor with neuroscientists who study the nature of memory, yet there is evidence for memory being embedded in our physiology—as expressed globally as "The First Principles of Physiology" (Torday and Rehan, 2009). More recently, Torday et al. have published extensively regarding the role of cell-cell communication in vertebrate evolution. Implicit in this reduction are the significance of memory and history in the process of cell-molecular evolution. This mechanism is consistent with claims that organ donor recipients have experienced personality changes, but the two concepts have not been collated up until now.

One documented example of cell donors affecting their recipients is Dolly the Sheep, the first cloned animal. Dolly was cloned by combining a cell taken from the mammary gland of a six-year-old Finn Dorset sheep with an egg cell taken from a Scottish Blackface sheep. Dolly died prematurely of pulmonary fibrosis. The cell used to transfer the genome to dolly was an adult mammary epithelial cell, which had not undergone the epigenetic changes conferred to an egg or sperm cell.

This way of thinking about consciousness as local and non-local is highly relevant to the work of philosophers David Chalmers and Andy Clark. Chalmers challenged our knowledge of consciousness with what he referred to as the 'hard problem', asking why it is that when we whack our thumb, we see 'red'? But based on body memory being located within cells, it is feasible that we see red when we feel pain because it is an atavistic recalling of when we first bled, having evolved a closed circulatory system. That is not homologous with our innate fear of snakes, spiders, the dark and lightening, for example. Clark has similarly challenged his colleagues by hypothesizing that the mind exists outside of the body, referring to it as the 'extended mind'. Chalmers and Clark characterized that as the 'informational extension' of the brain, like using cell phones and books, for example. It is hypothesized that they actually were expressing the non-local aspect of consciousness, transcending the body and referencing the gravitational force of the Cosmos.

Understanding the ontology and epistemology of life offers the opportunity to predict our future, and act accordingly. Up until now, we have been reasoning after the fact with the backdrop of the ambiguity of our origin, generating 'Just So Stories'. But now we have the option of acting based on a holistic narrative beginning from our initial conditions. In terms of basic science, there would be opportunity to use our knowledge of development and phylogeny prospectively for 'preventive medicine'. As for re-calibrating our worldview in order to be in sync with the environment, such initiatives as 'mindfulness', Quantum Consciousness, embodied mind can be expedited based on the predictive power of a model of consciousness that effectively integrates ontogeny and phylogeny. Time will tell.

The nature of consciousness has been questioned since the time of the Ancient Greek philosophers. Plato's 'cave' expressed a metaphoric consciousness as shadows; Aristotle thought consciousness was innate and of a 'higher order'; Descartes expressed the Mind-Body duality; Locke conflated consciousness with personal identity; William James saw consciousness as a continuum. To date, the problem has remained unanswered, yet the consensus is that it is the most important question facing us as a species.

Theories proposed by neuroscientists such as Gerald Edelman et al. and Antonio Damasio, and by philosophers like Daniel Dennett seek to explain consciousness in terms of the role of the brain. Many other neuroscientists, such as Crick and Christof Koch have explored the neural basis of consciousness without attempting to frame them as all-encompassing global theories. In tandem, computer scientists working in the field of artificial intelligence have pursued creating digital computer programs that can simulate or embody consciousness.

On July 7, 2012, eminent scientists from different branches of neuroscience convened at the University of Cambridge at the Francis Crick Memorial Conference, which deals with consciousness in humans and pre-linguistic consciousness in non-human animals. After the conference, they signed the 'Cambridge Declaration on Consciousness', summarizing the most important findings of the survey:

> "We decided to reach a consensus and make a statement directed to the public that is not scientific. It's obvious to everyone in this room that animals have consciousness, but it is not obvious to the rest of the world. It is not obvious to the rest of the Western world or the Far East. It is not obvious to the society."

> "Convergent evidence indicates that non-human animals ..., including all mammals and birds, and other creatures, ... have the necessary neural substrates of consciousness and the capacity to exhibit intentional behaviors."

The approach taken in this book, that consciousness is a derivative of physiology, would resolve the question of animal consciousness. Suffice to say that our central nervous system evolved from the skin of invertebrates, indicating that it originated in the physiology of the body. And for that matter, all living things—animal or vegetable—would be accounted for, given that both chloroplasts and mitochondria evolved from bacteria based on Symbiogenesis.

Physiology is holistic, having unitary characteristics because it is fractal, as first described by Mandelbrot—but how and why? It is self-similar at every scale, due to the underlying, integrative mechanisms generated by cellular ontogeny and phylogeny, constrained by the principle of homeostasis. The process is self-referential, referring all the way back in vertebrate phylogeny to its unicellular origin. But what are the integrating mechanisms that would account for the evolution of multicellular organisms from unicellular life? If they were known, they would provide fundamental insights to Natural Selection.

It has been argued that the First Principles of Physiology, identified by reverse-engineering the cell-cell signaling that gave form and function, are knowable since they are the mechanistic basis for ontogeny and phylogeny, which are one and the same process seen from the cellular perspective, occurring on different time-scales. That is to say, they are diachronic, or 'across space-time', given that the innate organizing principle of physiologic homeostasis is fractal. It is hypothesized that the epistatic balancing selection between calcium and lipid homeostasis was necessary for the initial conditions for eukaryotic evolution, triggering the process of vertebrate evolution, continuously perpetuating and embellishing it from unicellular organisms to metazoans in all phyla.

The germ lines (ectoderm, mesoderm, endoderm) monitor environmental changes and facilitate cellular evolution accordingly. These are the elements that fostered "fractal physiology". A classic misconception for the mechanism of evolution will be used to exemplify the difference between Darwinian random mutation and selection, on the one hand, and cellular-molecular evolution on the other. Peter Mitchell formulated his chemiosmotic hypothesis in 1961, that Adenosine Tri-Phosphate synthesized in respiring cells comes from the electrochemical gradient across the inner membranes of mitochondria by using the energy of NADH and FADH2 formed from the breakdown of energy-rich molecules such as glucose. Molecules such as glucose are metabolized to produce acetyl CoA as an energy-rich intermediate. The oxidation of acetyl CoA in the mitochondrial matrix is coupled to the reduction of carrier molecules such as NAD and FAD. The carriers pass electrons to the electron transport chain (ETC) of the inner mitochondrial membrane, passing them on to other proteins in the ETC. The energy available in the electrons is used to pump protons from the matrix across the inner mitochondrial membrane, storing energy as a transmembrane electrochemical gradient. The protons move back across the inner mitochondrial membrane through the enzyme ATP synthase. The flow of protons back into the matrix of the mitochondrion via ATP synthase provides enough energy for ADP to combine with inorganic phosphate to form ATP. The electrons and protons ultimately react with oxygen to form water.

Like Relativity Theory, the biology of multicellular organisms is also due to the interrelationships between space and time. In both cases, the Big Bang radiated out from its point of origin to give rise to signaling motifs, or Gene Regulatory Networks (GRNs) expressed as a function of their specific germlines, providing physical self-reference points for the physiologic internal environment to orient itself to the external physical environment. The genes take their "cues" from the interactions between the germline cells to generate form and function relative to the prevailing environmental conditions.

As a result, the organism can evolve in response to the ever-changing environmental conditions, the genes of the germline cells "remembering" previous iterations under which they (by definition) successfully mounted an adaptive response. Then, by recapitulating the germline-specific GRNs under newly-encountered conditions, they may form novel, phenotypically-adaptive structures and functions by recombining and permuting the old GRNs. This process is conventionally referred to as "emergent and contingent". It explains how and why the same GRN can be exploited to generate different phenotypes as a function of the history of the organism, both as ontogeny (short-term history) and phylogeny (short-term history), the germlines orienting and adapting the internal environment to the external environment by expressing specific genetic traits.

For example, seen from the traditional perspective, this phenomenon is described as pleiotropy. However, once we realize that this is actually the consequence of an on-going process, and not merely a chance occurrence, the causal relationships become evident, offering the opportunity to understand how and why form and function have evolved. Such interactive, cellular-molecular mechanistic pathways project both forward and backward in time and space, offering the opportunity to understand our unicellular origins, and the functional homologs that form the basis for the First Principles of Physiology.

In the beginning

Life likely began with the formation of liposomes through the agitation of lipids in water, bearing in mind that the moon separated from the Earth only 100 million years after the Earth was formed, offering billions of years for its effect on wave action to fashion life in this way. That interaction was chemiosmosis generating bioenergy, aiding and abetting negentropy, maintained by homeostasis. Endomembranes such as the nuclear envelope, endoplasmic reticulum, peroxisome and golgi apparatus formed intracellular compartments; compartmentation gave rise to the germ lines (ectoderm, mesoderm, endoderm) that monitor environmental changes and facilitate cellular evolution accordingly. These are the elements that fostered "fractal physiology".

Fractal properties of physiology seen as the evolution of calcium homeostasis

With the advent of the first cell, the biological world split into extra- and intra-cellular spaces that immediately began controlling the ion content of the cell cytoplasm. The first cell was defined by a membrane composed of ion-conducting pores and ion pumps in combination with chemiosmosis—

the movement of ions across a selectively-permeable membrane—as the source of energy to maintain entropy far from equilibrium. Failure of any of these components would have flooded the cytosol with calcium, testing the viability of the molecular machinery, potentially obviating the possibility of life.

The properties of the ion carriers were determined by the ion composition of the ocean that initially covered the Earth. There were few dissolved ions present at the time—sodium, chloride, magnesium, calcium, potassium and other trace ions. The concentration of Ca^{2+} was the most critical to biology because it denatures proteins, lipids and nucleic acids alike. At non-physiological concentrations, Ca^{2+} causes aggregation of proteins and nucleic acids, damaging lipid membranes, and precipitating phosphates. High intracellular levels of Ca^{2+} are incompatible with life; at all phylogenetic stages, from bacteria to eukaryotes, excess cytosolic Ca^{2+} is cytotoxic. Consequently, the first forms of life required effective Ca^{2+} control, maintaining intracellular Ca^{2+} at effectively low concentrations—around 100 nanomolar, which by comparison is approximately 1–2 thousand orders of magnitude lower than that in the extracellular milieu. Indeed, even the most primitive bacteria are endowed with plasmalemmal Ca^{2+} pumps (which are structurally similar to eukaryotic P-type Ca^{2+} pumps of either the PMCA or the SERCA types) and Ca^{2+}/H^+ and Na^+/Ca^{2+} exchangers.

High gradients of intra- and extra-cellular calcium ions have characterized life ever since its inception. The maintenance of such calcium ion concentrations requires substantial consumption of energy. Therefore, it would have been surprising if evolution had not led to adaptation of calcium ion homeostatic systems, whose initial functions were to protect the cell against massive, unrelenting calcium ion challenges, leading to ever-more complicated physiology.

Life 'authors' life

The value of the cellular-molecular approach to evolution is that it allows a view from the origins of life to its physiologic complications as one continuous arc. That view provides insight to the exaptational changes that have allowed for adaptation from earlier events in the organism's history that have likewise been exapted, succeeding in sustaining and perpetuating it. Based on that logic, one could hypothetically trace life all the way back to its origins. So how would that look? Hypothetically, the lipids in the primordial ocean that derived from the asteroid "snowballs" that formed them may have coalesced, and as the Sun warmed the Earth, such lipids would have liquefied, becoming more compliant. It has been postulated that those lipids formed micelles, or liposomes, which are semi-permeable protocells, offering a protective space for chemiosmosis and the reduction of their entropy, governed by homeostasis. Fast-forward that scenario and you can see how that facilitated many of the existential changes in the evolution of eukaryotic vertebrates, including the nuclear envelope, endomembranes like the endoplasmic reticulum, the cholesterol-containing phospholipid bilayer, peroxisomes, barriers like the skin, the gas exchange organs initiated by cholesterol as surfactant, and even the brain. This kind of phenomenology is usually referred to as self-organization or self-referencing (Maturana and Varela, 1972), but is not usually vertically integrated in this way.

Terminal Addition

Recent progress in our contemporary understanding of evolutionary development permits a re-framed appraisal of Terminal Addition—the 'practice' of adding newly-acquired traits onto the end of a series—as continuous historical cellular-environmental complementarity. Thereby, evolutionary terminal additions can be identified as episodic adjustments to cell-cell signaling.

Molecular cell embryology supports this, tracing development from the fertilized egg to the multicellular organism, and back again to the unicellular zygote over the course of the life cycle in an iterative manner from generation to generation. This perspective offers insight into the formation of the embryo, and how it grows, differentiates, and adapts to its environment. It is now known that the communication between cells in the developing conceptus is mediated by soluble growth factors, like the Spemann organizer, binding to its cognate receptors on neighboring cell-types to signal their presence and level of growth and differentiation. Target cells bind growth factors from their cell neighbors, triggering its developmental and resultant homeostatic programs. Such growth factors are highly conserved throughout evolution, and have been expressed in continuity from their unicellular origins forward across multicellular organisms, offering the opportunity to trace the molecular origins of vertebrates. Using this approach also leads to understanding how and why such patterns of cell-cell communication determine physiology, and why the breakdown in cell-cell communication leads to either adaptive strategies on the one hand, or maladaptive outcomes as another.

Terminal Addition as a consequence of evolved cell-cell signaling

It is asserted that evolutionary terminal additions can be identified as environmental induction of episodic adjustments to cell-cell signaling. This interaction yields cellular-molecular pathways that lead to differing developmental forms as a derivative manifestation of mutualistic/competitive cellular niche construction. This approach is not without precedent, and can be used as a functional genomic approach to evolution via Terminal Addition. Horowitz formulated a similar approach to the evolution of biochemical pathways by assuming a retrograde mode of evolution. In contrast to that, the cellular-molecular paracrine mechanism underpinned by a series of ligand-receptor interactions that evolved in response to a series of external (atmospheric oxygen, stretch) and internal (metabolic demand, tissue oxygenation, alveolar surface tension, blood pressure) selection pressures would have caused the evolution of the homeostatic mechanisms that determined those biosynthetic pathways from phenotypes to genes, i.e., cell auto-engineering. Subsequent selection pressure for such ligand-receptor-mediated gene regulatory networks would have generated both evolutionary stability and novelty through such well-known mechanisms as gene duplication, gene mutation, redundancy, alternative pathways, compensatory mechanisms, and balancing selection pressures. These genetic modifications were manifested by the structural and functional changes in the gas exchanger, primarily by the thinning of the blood-gas barrier in conjunction with adaptive phylogenetic changes in the composition of the surfactant. The reverse engineering of these phenotypic changes in the blood-gas barrier form the basis for a molecular genetic approach to lung evolution. It should be noted that this path provides a novel emergent and contingent mechanism for evolution. More importantly, this model of cellular-molecular evolution predicts the evolution of other physiologic mechanisms by integrating reproduction into the selection pressure processed specifically, at each proximal step in the retrograde evolution of surfactant, its physiologic roles, either as a newly evolved step or as a functionally interrelated aspect of integrated physiology, would have been constrained by the immediate and related mechanisms that prepare the embryo for its homeostatic adaptation to extrauterine life.

For example, stretch-regulated PTHrP signaling and surfactant production interrelate functionally (and genomically) with their complementary roles in bone development, skin maturation, the birth process, and glomerular physiology. Going one step further, by tracing its evolution backwards, this same cellular-molecular processing from proximate to ultimate physiologic characteristics has been canonically repeated, perhaps rooted in its unicellular origins the cell-molecular mechanism of lung evolution based on the evolution of the surfactant dovetails with fundamental mechanisms of membrane evolution put forward by Konrad Bloch, and by Thomas Cavalier-Smith. Bloch demonstrated that cholesterol evolved in response to the rise of oxygen in the atmosphere, speculating that its biologic advantage was due to the reduced fluidity, or increased microviscosity resulting from the addition of cholesterol to the cell membrane phospholipid bilayer. The discovery of hopanoid triterpene derivatives in some prokaryotes in the form of "molecular fossils" of ancient times, has led to the suggestion that these relatively rigid, anaerobically evolved, amphiphilic molecules play a membrane reinforcing role in some prokaryotes similar to that exhibited by aerobically evolved sterols such as cholesterol in eukaryotes. In turn, the advent of cholesterol might have constrained cytosis. Therefore, there is a cell-molecular continuum from the evolution of cholesterol for the compliance of the plasma membrane of unicellular eukaryotes, to endocytosis/exocytosis in eukaryotes, to the efficient functioning of the swim bladder resulting from the secretion of cholesterol as a lubricant, to lung surfactant reducing surface activity.

The characterization of the above cellular-molecular events as Terminal Addition is corroborated by the epithelial-mesenchymal interactions involved in the development and phylogeny of the lung alveolus, beginning with Parathyroid Hormone-related Hormone (PTHrP) production and secretion by the alveolar epithelium binding to its receptor on the mesenchyme, triggering a cyclic Adenosine Monophosphate cascade that causes differentiation of the lung myofibroblast into the lipofibroblast. In turn, the lipofibroblast produces and secretes leptin, which binds to its cell surface receptor on the epithelial type II cell, stimulating the production of lung surfactant. Leptin also stimulates the transit of fatty acid from the lipofibroblast to the alveolar type II cell, mediated by the production of Prostaglandin E_2 by the alveolar type II cell. In the context of Terminal Addition, each of these intermediary steps in lung alveolar development comply with the phylogeny of the alveolus from fish to amphibians, reptiles, mammals and birds.

As for experimental evidence for the causal evolutionary nature of this mechanism, when cholesterol synthesis by the alveolar type II cell is deleted, it decreases the surface-tension reducing capacity of the surfactant, resulting in stress on the alveolus. To compensate for this effective surfactant deficiency, the alveolus 'recapitulates' the evolutionary formation of lipofibroblasts during development to compensate for the loss of surface-activity and prevent alveolar collapse. Thus, the steps in alveolar evolution from the swim bladder of fish to the advent of the lipofibroblast in preventing oxidant injury is reprised both developmentally and phylogenetically as originally described by Haeckel. It is an important outgrowth of this in-depth understanding of the underlying cellular pathways that a group of bioactive molecules appear to have similar developmental roles in some insects, crustaceans and chelicerates. This provides evidence of homologies of linked molecular action and signaling pathways across these arthropods that has deeper evolutionary meaning than morphologic overlap.

Epigenetic impacts and terminal addition

Terminal additions are mediated by epigenetic means as a mechanism for accommodating environmental stresses, first at the somatic level and then through the germline; but in order to understand the significance of Terminal Addition, it is first necessary to clarify some aspects of evolutionary development that have consistently been misconstrued. Despite any conventional visual appraisal, the phenotype is a means by which macro-organisms acquire epigenetic experiences for the perpetuation of the dominant eukaryotic unicellular form. For eukaryotes, the means by which terminal additions are added to holobionts through self-referential consensual cellular collaborations as niche construction is the phenotype. The phenotype is therefore purposed towards environmental exploration that returns from phenotype to be indirectly experienced through obligatory recapitulation to the unicellular form. The unicellular zygote adjudicates epigenetic impacts, and therefore has a crucial role in the addition of appropriate forms of Terminal Addition or their heritable exclusion.

On the evolutionary significance of phantom limb syndrome

The phenomenon of Phantom Limbs makes one wonder what selection advantage it offers. Upon further reflection, it is consistent with the concept of the "phenotype as agent" as the primary purpose of the organism, functioning to obtain epigenetic marks over the course of its life cycle. Pseudopods evolved during the unicellular stage of vertebrate evolution along with metabolism and oxygenation, particularly when cholesterol appeared in the cell membrane, and as such are functionally interrelated with all of the other biologic traits that evolved over the subsequent course of vertebrate evolution. Loss of a limb does not eliminate the need to collect epigenetic marks in conjunction with all of the other physiologic relationships that must be sustained if the organism is to remain evolutionarily competitive with other organisms. The exception proves the rule yet again.

The phenomenon of sensing a severed limb is conventionally only thought of in the context of the traumatic loss, as would be expected given the circumstances. However, it may be construed as a manifestation of the "phenotype as agent," the organism evolutionarily mandated to obtain epigenetic marks, given that its evolved state is the net result of structural–functional changes over the history of the organism, beginning in the unicellular state. Since limb development is an essential aspect of that process, its emergence is a result of earlier events, beginning with the acquisition of cholesterol in the cell membrane of unicellular eukaryotes, promoting metabolism, oxygenation, and most importantly in the context of phantom limb sensation, locomotion. The latter originated as increased cytoplasmic streaming resulting from the appearance of cholesterol in the cell membrane, over time evolving as limbs in multicellular organisms, linked to metabolism and respiration by cholesterol from the inception of vertebrate evolution. That connection is highlighted because it provides the rationale for the etiology of phantom limb sensation since all of these properties are interconnected and must act in unison to facilitate the collection of epigenetic marks in order for the organism to faithfully monitor its environment and inform itself of any changes via the germ cells, zygote, embryo, and offspring in the next generation.

Background to phantom limb sensation

The predominant hypothesis for phantom limbs has been that it is due to the inflammation of the severed nerve endings. Given the absence of the limb, this phenomenon was strictly thought of as the brain perceiving pain. However, Ronald Melzack disputed this hypothesis, instead proposing the neuromatrix hypothesis, that the body is composed of a wide network of interconnecting neural networks. Pons et al. subsequently showed experimentally that the somatosensory cortex of the brain reorganizes after loss of sensory input in macaque monkeys. Ramachandran and Hirstein, in turn, hypothesized that phantom limb sensations in humans could be due to the sensorimotor cortex, located in the postcentral gyrus, which receives input from the limbs and body. He and his associates demonstrated this hypothesis by showing that stimulation of different parts of the face caused perceptions of being touched on different parts of the missing limb. Flor et al. showed that the pain of limb loss was the result of the cortical reorganization, whereas Knecht et al. concluded that there was no relationship between referred sensations and cortical reorganization within the primary cortical areas. Flor has also found that non-painful referred sensations correlate with a wide neural network beyond the primary cortical areas. Despite such active research into the neural mechanisms of phantom limb sensation, there is no consensus as to the cause.

Relevance of phantom limb sensation to Terminal Addition

Terminal Addition is a historically-based mechanism for cellular–environmental complementarity, mediated by cell–cell interactions and their downstream signaling cascades. Such relationships refer all the way back to the unicellular origins of the organism, and as such have formed both linear and nonlinear network interconnections with collateral structures and functions in ways that are consistent with the evolution of the organism. As such, any interference with such networks disrupts the agency of the phenotype, endangering the ability of the organism to fulfill its responsibility as communicator of epigenetics. Seen in this context, phantom limb sensation makes sense.

Phantom limb sensation as quantum entangled non-locality

The concept of non-localization has been discussed at length by Bohm and Hiley. They bring out the fact that the essential new quality implied by the quantum theory is non-locality; i.e., that a system cannot be analyzed into parts whose basic properties do not depend on the state of the whole system, showing that this approach implies a new universal type of description, in which the standard or canonical form is always supersystem-system-subsystem, leading to the radically new notion of unbroken wholeness of the entire Universe.

Biology ascribes to the same description. It is not apparent when seen from a synchronic descriptive vantage-point, but when understood from a diachronic perspective, transcending space and time, it can be understood in the same terms used by Bohm and Hiley. This way of thinking about biology in cellular–molecular terms is exemplified by recalibrating of pleiotropy. In contrast to the stochastic way of conventionally thinking about pleiotropy as the random expression of genes throughout the organism to generate more than one distinct phenotypic trait, it is actually a deterministic consequence of the evolution of complex physiology from the unicellular state. Pleiotropisms emerge through recombinations and permutations of cell–cell communication established during meiosis based on the history of the organism, both developmentally and phylogenetically, in service to the future existential needs of the organism. Functional homologies ranging from the lung to the kidney, skin, brain, thyroid, and pituitary exemplify the evolutionary mechanistic strategy of pleiotropy. The power of this perspective is exemplified by the resolution, for example, of evolutionary gradualism and punctuated equilibrium in much the same way that Niels Bohr resolved the paradoxical duality of light as complementarity. Hence, seen in this way, biology and physics are both non-localized, acting at all scales to form and maintain their integrated entirety.

Symbiogenesis meets quantum entanglement

It is one thing to trace the evolution of physiologic traits, but the ontogeny and phylogeny beg the question as to where they originated? The methodology used to reverse-engineer that was Bayesian in nature, homing in on known emergences over the course of vertebrate evolution, beginning with the significance of lipids for the progressive decrease in the diameter of the alveoli in order to increase oxygenation. To be clear, the only way that oxygenation could be enhanced physiologically in order to facilitate metabolism on land was by increasing the blood volume-to-surface area ratio. But in order to accomplish that lung surfactant had to evolve progressively in tandem in order to prevent alveolar collapse due to the inherent surface tension of the microscopic water layer lining the interior surfaces of the alveoli. That step-wise process emanated from the evolution of the lipofibroblast in the subepithelial connective tissue [Torday and Rehan, 2016], acting to protect the alveoli from the rising levels of oxygen in the environment. And superimposed on those critical steps in lung evolution was the evolution of the Parathyroid Hormone-related Protein (PTHrP) signaling for super-regulation of lung surfactant, originating from the duplication of the PTHrP Receptor gene during the transition from water to land.

Preceding that cascade of cellular-molecular events was the evolution of the peroxisome, which De Duve hypothesized was in response to endoplasmic reticulum stress causing the flooding of the cytoplasm with reactive oxygen species, the lipid content of the peroxisomes being released to buffer the latter and prevent toxicity.

And prior to that, Konrad Bloch discovered the cholesterol biosynthetic pathway, hypothesizing that in order to do so there had to have been enough oxygen in the atmosphere, given that it takes 11 atoms of oxygen to produce one molecule of cholesterol......but that's reasoning after the fact. Seen from the perspective of the unicell, cholesterol had a coordinated physicochemical effect on cellular evolution, thinning the cell membrane, enhancing oxygenation and metabolism.

Now, turning the 'equation' around, it was Symbiogenesis—the assimilation of factors in the environment that posed existential threats—that was the driver for eukaryotic cell evolution. But that begs the question as to where and how Symbiogenesis evolved? Since it essentially ensures homeostatic balance for the cell, and the consensus is that Quantum Mechanics is at the source of all things Cosmos, including life, it has been hypothesized that Quantum Entanglement is the homologue of Symbiogenesis, maintaining the homeostatic balance of particles entering the cell via the semi-permeable lipid membrane. If so, the salient feature of QE is local/non-local properties, providing a rationale for the nature of consciousness as both local and non-local. That is to say, local consciousness is the aggregate of physiology [Torday, 2020], whereas non-local consciousness is 'cosmic', the common denominator for the two states being the Laws of Nature.

Experimental evidence for the local/non-local nature of consciousness lies in the result of putting differentiated cells in microgravity [Torday, 2003]. Under such conditions, the cell loses its phenotypic identity, yet it has the capacity to 'remember' it because when put back in unit gravity the cell regains its phenotype. If local/non-local consciousness is a binary, it is hypothesized that the cell shifts to its cosmic state of consciousness, but remains in contact with its local consciousness enough to be able to rebalance in unit gravity. This way of understanding the mechanism of consciousness would provide explanations for Near Death Experiences, Out of Body Experiences, Maslow's Peak Experiences, the Runner's High, the meditative state, mindfulness and dreams. The recognition of these properties as part and parcel of the continuum of consciousness has the potential to expand consciousness towards 'super' consciousness.

References cited

Balasubramanian, P. and Longo, V.D. 2016. Growth factors, aging and age-related diseases. Growth Horm. IGF Res. 28: 66–68.

Maturana, H.R. and Varela, F.J. 1972. Autopoiesis and Cognition: the Realization of the Living. Boston Studies in the Philosophy and History of Science. Reidl Publishing Company, Boston.

Pinheiro, P.L., Cardoso, J.C., Power, D.M. and Canário, A.V. 2012. Functional characterization and evolution of PTH/PTHrP receptors: insights from the chicken. BMC Evol. Biol. 6: 110.

Sagan, L. 1967. On the origin of mitosing cells. J. Theor. Biol. 14: 255–274.

Torday, J.S. 2003. Parathyroid hormone-related protein is a gravisensor in lung and bone cell biology. Adv. Space Res. 32: 1569–1576.

Torday, J.S. and Rehan, V.K. 2009. Lung evolution as a cipher for physiology. Physiol. Genomics 38: 1–6.

Torday, J.S. 2015. Homeostasis as the Mechanism of Evolution. Biology (Basel) 4: 573–590.

Torday, J.S. and Rehan, V.K. 2016. On the evolution of the pulmonary alveolar lipofibroblast. Exp. Cell Res. 340: 215–219.

Torday, J.S. 2020. Consciousness, Redux. Med. Hypotheses 140: 109674.

Zhang, X. and Ho, SM. 2011. Epigenetics meets endocrinology. J. Mol. Endocrinol. 46: R11–R32.

Chapter 15
Embodied Quantum Entanglement

"If you think you understand quantum mechanics, you don't understand quantum mechanics."

Richard Feynman

Introduction

In a stage of human history when the line between truth and fallacy, fact and fiction are being blurred, our ignorance is due to lack of understanding of our ontology and epistemology. Then in 2003 Thomas Berry recited a new story, recounting the work of Teilhard de Chardin. This chapter is in honor of that plea.

We have unfortunately misconstrued our understanding of consciousness, our own physiology as the source of that 'awareness' (Torday and Rehan, 2017). Instead of looking within ourselves, beyond the brain, William of Ockam remonstrating to 'keep it simple', we have looked outward using technologies to reduce biology and physics to their smallest components, concluding that both are 'probabilistic'. It is maintained that this misguided conclusion is the result of the Scientific Method, relying on binary decisions, or 'yes/no' hypotheses based on experimentation, when in reality life is 'holistic'. But there is a way to rectify this systematic error through a deep understanding of what life 'means' in order to further advance our knowledge of who and what we are.

Life initialized from lipid molecules derived from deep space floating on the surface of the primordial ocean that covered the Earth's surface. Such lipid molecules stand perpendicularly to the water's surface under the force of gravity, with their negatively charged hydrophilic ends facing downward into the water, their hydrophobically-charged ends pointing skyward. Once a critical negative charge has been amassed, it neutralizes the Van Der Waals force for the surface tension of water, causing a quantum transition from single lipid molecules to micelles, or primitive cells, representing the origin of cell physiology.

The First Principles of Physiology

Elsewhere, it has been hypothesized that there are 'First Principles of Physiology' [Figure 1] based on self-organizing mapping of ontogeny and phylogeny, begging the question as to where such 'principles' emerged from? The three proposed principles are negative entropy, chemiosmosis, and homeostasis. Based on experimental evidence for the fundamental role of gravity in the formation of the first cell from lipid molecules (Torday, 2003), the energy generated by gravity would have facilitated the Quantum Entanglement of particles within the cell, accounting for negative entropy within the cell. All of the above would have been under the control of homeostasis (Cannon, 1932).

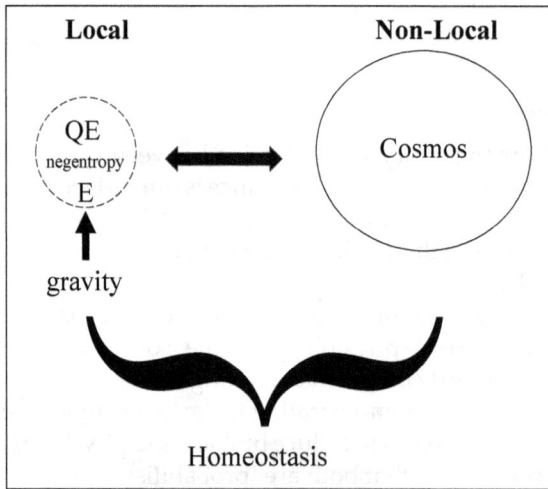

Figure 1. Origin of The First Principles of Physiology. Gravity imparted the energy needed for Quantum Entanglement of particles entering the cell, allowing for negentropy. That constitutes "local" consciousness, with reference to the non-local consciousness of the Cosmos. The local and non-local consciousness are held in balance by homeostasis.

On the origin of Symbiogenesis

The step-by-step filling of the micelle with traits assimilated from the environment is referred to as Symbiogenesis (Sagan, 1967) [Figure 2]. That raises the question as to what the pre-adaptation for Symbiogenesis was, maintaining homeostasis in the face of existential threats? When gravity impinges on a curved surface [Figure 2], it generates energy

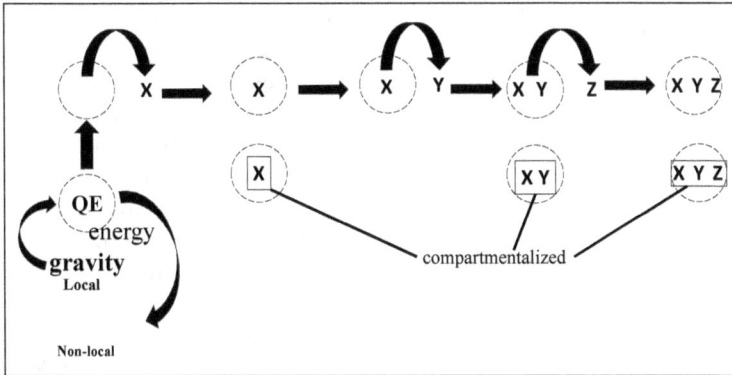

Figure 2. The Effects of Local versus Non-Local Gravity. Symbiogenesis is depicted as the sequence of assimilations of environmental factors x, y and z. Local gravity (Earth) formed cells (lower left). In turn, local gravity referenced non-local gravity in the Cosmos based on Quantum Entanglement (QE). The symbiogenic assimilation of existential threats to the organism (X Y Z, etc.), formed the basis for physiology as consciousness, linking the local of the Explicate Order with the non-local of the Implicate Order, or the Cosmos as Consciousness.

based on Einstein's Field Theory (Einstein, 1961). That energy would have sustained Quantum Entanglement (QE) operating within the micelle, maintaining particles entering the micelle across its semi-permeable membrane. And based on QE, the local effect of Earth's gravity on the micelle would have been 'counter-balanced' homeostatically by the non-local gravity in the Cosmos. Seen from this perspective, Symbiogenesis is the exapted mechanism for maintaining the homeostatic balance [see Figure 2] between the local and non-local in an ever-expanding Universe.

The cell can be seen as operating based upon both Newtonian and Quantum Mechanical principles, acting as the unifying agent for both, as observer and observed. The capacity to unify Quantum and 'classical' physics is referred to as coherence, which has been disputed for biology in the past. Yet the way in which Quantum Entanglement forms the basis for Symbiogenesis now explains that property. This is the reason why life functions at both the terrestrial and Cosmic levels, as expressed by consciousness, behaving at both the local and non-local levels.

The effect of microgravity causing the loss of phenotypic identity reveals the Two Levels of Consciousness experimentally. By exposing lung or bone cells to microgravity, it was observed that they lose their differentiated phenotypes (Torday, 2003), but their phenotypic identity can be restored readily by re-exposing the cells to gravity, indicating the non-damaging nature of the effect. Moreover, based on QE conferring locality/non-locality, absent the local effect of gravity, the cell in microgravity would, by definition, revert to the non-local cosmologic state. In support of this in vitro experiment, bones from rats flown in deep space for two weeks (STS-58, SLS-2, courtesy of the National Aeronautics and Space Administration) were assayed for their Parathyroid Hormone-related Protein (PTHrP) mRNA content, PTHrP acting as a paracrine integrator of physiology. It was found that the amount of PTHrP mRNA was significantly lower in the space-based bones than in bones from littermate control rats that remained on Earth. The consequence of decreased PTHrP expression in bone is loss of calcium, resulting in osteoporosis, a well-known consequence of space flight (Cappellesso et al., 2015).

Equivalent results were obtained when yeast, unicellular eukaryotes, were exposed to microgravity (Purevdorj-Gage et al., 2006). The experimentally-treated yeast were unable to conduct a calcium flux or reproduce. The inability to generate a calcium flow rendered them unable to respond to environmental stimuli. In the aggregate, microgravity-exposed yeast cells had lost their phenotypic identity just like the differentiated lung and bone cells had, supporting the observed role of microgravity on eukaryotic phenotypic identity.

Consciousness, local and non-local

Elsewhere, it has been proposed that physiology is the aggregate of our consciousness. When consciousness is lost under such circumstances as Near Death Experience, Out of Body Experience, Maslow Peak Experience, meditation, or the proverbially Runner's High, the local effect of gravity on physiology as ego is diminished or lost, resulting in the organism defaulting to the non-local gravity of the Cosmos. Likewise, as we age or senesce, there is a loss of cell-cell signaling due to decline in growth factor receptors, leading to a gradual loss in the local effect of gravity, defaulting to the non-local gravity of the Cosmos.

Although definitive evidence for the homeostatic balance between the local and non-local gravitational effects on consciousness is not forthcoming, there is circumstantial evidence like the recapitulation of brain evolution among patients undergoing general anesthesia (Mashour and Alkire, 2013), or the Libet Experiment (Libet et al., 1983), in which subjects undergoing brain monitoring were administered an electrical shock. It was observed that the stimulation of brain activity preceded the physical reaction by 300 milliseconds, inferring that the mind knows how to react before the body does.

Even the 'Double-slit' experiment, casting a light beam on a template with two slits in it, generating two patterns when observed indicates the dual nature of consciousness. Namely, when don't look, the light beam becomes one continuous path. It has been suggested that the particles displayed on the wall are indicative of the digitized cell membrane, the continuum of the light wave being indicative of our physiology, 'remembering' our history.

"Body memory" as consciousness?

Currently, based on cellular memory, body cells contain information about our personalities, likes, dislikes, and our overall history unrelated to DNA or the brain. In this context, the recent discovery that the endocrine system is under epigenetic control is a game changer, because it infers that the effects of the hormones that control our behavior can be directly inherited from the environment.

Epigenetic Inheritance occurs when factors in the environment are perceived as existential threats, and are assimilated by the egg and sperm as 'epigenetic marks'. Such marks modify the development of the offspring in order to adapt. This process is consistent with the organism being an 'agent' for detecting changes in its environment. This mechanism is also consistent with inheritance being described as Terminal Addition, newly evolved traits being added on the end of a sequence of evolved traits, since such traits are determined by cell-cell signaling. That is to say, adding a new trait to the middle or to the beginning of such series of evolved traits would be highly inefficient, given that the mechanism entails upstream effects of 'second messengers' that affect growth and differentiation.

In the book "Molecules of Emotion", Candace Pert expressed the idea that "the mind is not just in the brain, but also exists throughout the body." She claimed that "The mind and body communicate with each other through chemicals known as peptides". These peptides are found in the brain as well as in the stomach, muscles, all throughout our major organs and tissues. She believed that memory could be accessed from anywhere in the peptide/hormone/receptor network. For example, a food memory might be linked to the pancreas or liver, and such associations could be transferred from one person to another.

Since evidence for such claims has yet to be produced, Pert's notions have not found acceptance with neuroscientists who study memory. More recently, Torday et al. have published extensively (Torday and Rehan, 2012; Torday and Rehan, 2017; Torday and Miller, 2020; Guex et al., 2020; Torday et al., 2019) regarding the role of cell-cell communication in vertebrate evolution. Explicit to that reduction is the significance of memory and history in the process of cell-molecular evolution. That mechanism is consistent with claims that organ donor recipients have experienced personality changes (Pearsall et al., 2000), but the two concepts have not been melded together until now.

One high-profile example of cells from donors affecting their recipients is Dolly the Sheep, the first cloned animal. She was cloned from a cell obtained from the mammary gland of a six-year-old Finn Dorset sheep and an egg cell taken from a Scottish Blackface sheep. She died prematurely of lung fibrosis. The cell used to transfer the genome to dolly was an adult mammary epithelial cell, which would not have undergone the epigenetic changes conferred to an egg or sperm cell.

This way of thinking about consciousness as local and non-local is consistent with the work of David Chalmers and Andy Clark. Chalmers challenged his fellow philosophers with the 'hard problem', asking why when we whack our thumb, we see 'red'? Based on body memory being within cells, it is feasible that we see red when we feel pain because it is a recalling of when we first bled in association with feeling pain, not unlike our innate fear of snakes, spiders, the dark and lightening. Clark and Chalmers have also challenged their colleagues by hypothesizing that the mind extends beyond the body (Clark and Chalmers, 1998), referring to that property as the 'extended mind'. They characterized that as the 'informational extension' of the brain using cell phones and books as examples. I propose that they actually were expressing the non-local aspect of consciousness, transcending the body and referencing the gravitational force of the Cosmos.

Discussion

Understanding the ontology and epistemology of life offers the opportunity to predict our future, and behave accordingly. Up until now, we have been reasoning after the fact, coping with the ambiguity of our origin by telling ourselves 'Just So Stories'. But now, with the insight like those referenced in this article, we have the option of acting based on a narrative beginning from our initial conditions, our knowledge-gathering consistent with epigenetics. That is the hope expressed in this chapter. In terms of basic science, there would be opportunity to use our knowledge of development and phylogeny prospectively for 'preventive medicine'; and as for re-calibrating our worldview in order to sync with the environment, such initiatives as 'mindfulness', Quantum Consciousness, and embodied mind can be expedited based on the predictive power of a model of consciousness that effectively integrates ontogeny and phylogeny. Only time will tell.

The nature of consciousness has been formally pursued since the time of the Ancient Greek philosophers. Plato expressed the concept of a 'cave' in which were stored archetypes (Plato, 2007). Aristotle thought that consciousness was intrinsic and higher order (2011). In more recent times, Descartes expressed the Mind-Body duality (Cunning, 2014), whereas Locke equated consciousness with personal identity (Nimbalkar, 2011). William James saw consciousness as a continuum (1890). The problem has remained unanswered to this day, yet the consensus is that it is the most important question facing us as a species.

Theories proposed by neuroscientists such as Gerald Edelman et al. (2011) and Antonio Damasio (2000) and by philosophers such as Daniel Dennett (1992) seek to explain consciousness in terms of neural events occurring within the brain. Many other neuroscientists, such as Crick and Christof Koch (1990) have explored the neural basis of consciousness without attempting to frame all-encompassing global theories. At the same time, computer scientists working in the field of artificial intelligence have pursued the goal of creating digital computer programs that can simulate or embody consciousness (Sun and Franklin, 2007).

On July 7, 2012, eminent scientists from different branches of neuroscience gathered at the University of Cambridge to celebrate the Francis Crick Memorial Conference, which deals with consciousness in humans and pre-linguistic consciousness in nonhuman animals. After the conference, they signed the 'Cambridge Declaration on Consciousness' in the presence of Stephen Hawking, which summarizes the most important findings of the survey:

"We decided to reach a consensus and make a statement directed to the public that is not scientific. It's obvious to everyone in this room that animals have consciousness, but it is not obvious to the rest of the world. It is not obvious to the rest of the Western world or the Far East. It is not obvious to the society."

"Convergent evidence indicates that non-human animals ..., including all mammals and birds, and other creatures, ... have the necessary neural substrates of consciousness and the capacity to exhibit intentional behaviors."

The approach taken in this article, that consciousness is a derivative of physiology, would resolve the question of animal consciousness. And suffice to say that our central nervous system evolved from the skin of invertebrates (Holland, 2000), indicating that it originated in the physiologic makeup of the body. And for that matter, all living things, animal or vegetable would be accounted for, given that both chloroplasts (Cavalier-Smith, 2002) and mitochondria (Sagan, 1967) evolved from bacteria based on Symbiogenesis.

References cited

Aristotle. 2011. The Philosophy of Aristotle. Signet Classics, New York.

Balasubramanian, P. and Longo, V.D. 2016. Growth factors, aging and age-related diseases. Growth Horm. IGF Res. 28: 66–68.

Berry, T. 2003. Teilhard in the 21st Century: The Emerging Spirit of Earth. Orbis Books, Maryknoll.

Campbell, K.H., McWhir, J., Ritchie, W.A. and Wilmut, I. 1996. Sheep cloned by nuclear transfer from a cultured cell line. Nature 380: 64–66.

Cannon, W.B. 1932. The Wisdom of the Body. Norton, New York.

Cappellesso, R., Nicole, L., Guido, A. and Pizzol, D. 2015. Spaceflight osteoporosis: current state and future perspective. Endocr. Regul. 49: 231–239.

Cavalier-Smith, T. 2002. Chloroplast evolution: secondary symbiogenesis and multiple losses. Curr. Biol. 12: R62–R64.

Chalmers, D. 1995. Facing up to the hard problem of consciousness. J. Conscious. Stud. 2: 200–219.

Clark, A. and Chalmers, D.J. 1998. The extended mind. Analysis 58: 7–19.

Crick, F.C. and Koch, C. 1990. Towards a neurological theory of consciousness. Seminars Neurosci. 2: 263–275.

Cunning, D. 2014. The Cambridge Companion to Descartes' Meditations. Cambridge University Press, Cambridge.

Damasio, A. 1999. The Feeling of What Happens: Body and Emotion in the Making of Consciousness. Harcourt Press, Boston.

Damasio, A. 2000. The Feeling of What Happens. Mariner Books, Boston.

Darwin, C. 1859. On the Origin of Species. John Murray, London.

Dennett, D.C. 1992. Consciousness Explained. Back Bay Books, Boston.

Edelman, G.M., Gally, J.A. and Baars, B.J. 2011. Biology of consciousness. Front. Psychol. 2: 4.

Einstein, A. 1961. Relativity. The Special and General Theory. Crown Publishing, New York.

Feynman, R.P. 1951. Berkeley Symposium on Mathematical Statistics and Probability. 2: 533–541.

Guex, J., Torday, J.S. and Miller, W.B. Jr. 2020. Morphogenesis, Environmental Stress and Reverse Evolution. Springer Nature, Switzerland.

Holland, N.D. 2003. Early central nervous system evolution: an era of skin brains? Nat. Rev. Neurosci. 4: 617–27.

James, W. 1890. The Principles of Psychology. Macmillan, London.

Libet, B., Gleason, C.A., Wright, E.W. and Pearl, D.K. 1983. Time of conscious intention to act in relation to onset of cerebral activity (readiness-potential). The unconscious initiation of a freely voluntary act. Brain 106: 623–642.

Mashour, G.A. and Alkire, M.T. 2013. Evolution of consciousness: phylogeny, ontogeny, and emergence from general anesthesia. Proc. Natl. Acad. Sci. U S A 2: 10357–10364.

Mitchell, P. 1961. Coupling of phosphorylation to electron and hydrogen transfer by a chemi-osmotic type of mechanism. Nature 191: 144–148.

Nimbalkar, N. 2011. John Locke on personal identity. Mens Sana Monogr. 9: 268–275.

Pearsall, P., Schwartz, G.E. and Russek, L.G. 2000. Changes in heart transplant recipients that parallel the personalities of their donors. Integr. Med. 2: 65–72.

Pert, C. 1999. Molecules of Emotion. Simon and Schuster, New York.

Plato. 2007. The Republic. Penguin Classics, London.

Purevdorj-Gage, B., Sheehan, K.B. and Hyman, L.E. 2006. Effects of low-shear modeled microgravity on cell function, gene expression, and phenotype in *Saccharomyces cerevisiae*. Appl. Environ. Microbiol. 72: 4569–4575.

Sagan, L. 1967. On the origin of mitosing cells. J. Theor. Biol. 14: 255–274.

Salama, F. 2008. PAH's in Astronomy—a Review. Organic Matter in Space. Proceedings I.A.U. Symposium 251: 357–365.

Schrodinger, E. 1944. What is Life? MacMillan, New York.

Skinner, M.K. and Nilsson, E.E. 2021. Role of environmentally induced epigenetic transgenerational inheritance in evolutionary biology: Unified Evolution Theory. Environ. Epigenet. 7: dvab012.

Sun, R. and Franklin, S. 2007. Computational Models of Consciousness: A Taxonomy and Some Examples. The Cambridge Handbook of Consciousness. Cambridge University Press, Cambridge.

Tang, Y.Y., Hölzel, B. and Posner, M. 2015. The neuroscience of mindfulness meditation. Nat. Rev. Neurosci. 16: 213–225.

Torday, J.S. 2003. Parathyroid hormone-related protein is a gravisensor in lung and bone cell biology. Adv. Space Res. 32: 1569–1576.

Torday, J.S. and Rehan, V.K. 2009. Lung evolution as a cipher for physiology. Physiol. Genomics 38: 1–6.

Torday, J.S. and Rehan, V.K. 2012. Evolutionary Biology, Cell-Cell Communication and Complex Disease. Wiley, Hoboken.

Torday, J.S. 2015. Homeostasis as the mechanism of evolution. Biology (Basel) 4: 573–590.

Torday, J.S. and Miller, W.B. 2016. Phenotype as agent for epigenetic inheritance. Biology (Basel) 5: 30.

Torday, J.S. and Miller, W.B. Jr. 2017. The resolution of ambiguity as the basis for life: A cellular bridge between Western reductionism and Eastern holism. Prog. Biophys. Mol. Biol. 131: 288–297.

Torday, J.S. and Rehan, V.K. 2017. Evolution, the Logic of Biology. Wiley, Hoboken.

Torday, J.S. 2018. Quantum Mechanics predicts evolutionary biology. Prog. Biophys. Mol. Biol. 135: 11–15.

Torday, J.S. 2019. The Singularity of nature. Prog. Biophys. Mol. Biol. 142: 23–31.

Torday, J.S. 2020. Consciousness, Redux. Med. Hypotheses 140: 109674.

Torday, J.S. and Miller, W.B. Jr. 2020. Cellular-Molecular Mechanisms in Epigenetic Evolutionary Biology. Springer Nature, Switzerland.

Varela, F.J., Rosch, E. and Thompson, E. 2017. The Embodied Mind. MIT Press, Cambridge.

Wigner, E. 1961. Remarks on the Mind-Body Question. The Scientist Speculates. Heinemann, London.

Young, T. 1804. The Bakerian Lecture. Experiments and calculation relative to physical optics. Philos. Trans. R. Soc. Lond. 94: 1–16.

Zhang, X. and Ho, S.M. 2011. Epigenetics meets endocrinology. J. Mol. Endocrinol. 46: R11–R32.

Zurek, W. 2002. Decoherence and the transition from Quantum to classical—Revisited. Los Alamos Science 27: 2–25.

Chapter 16
If Music Be the Food of Love, Play On....

Introduction

This title is one of the most well-known opening lines in all of English literature. It raises the question as to how and why music affects consciousness? I am of the opinion that only the Scientific Method can elevate our level of consciousness, leaving open the question as to the role of music in reinforcing our consciousness.

In order to come up to speed on the role of music for Man, I read Michael Spitzer's "The Musical Human". This is a thoroughly engaging work that allows one to understand music's varied functions in human endeavors, but does not address its role in consciousness. I will attempt to do so, as follows.

Let's start from the premise that the cell 'rectifies' noise in its environment using the logic of its existence. But that raises the question as to how and why it does so? The working hypothesis is that the cell is an agent for detecting change in the environment in order to maintain homeostatic equipoise, given that the environment is ever-changing due to the on-going expansion of the Cosmos. This is consistent with the larger hypothesis that the cell's role is to communicate with the environment and with other cells. When the noise is extreme, it leads to loss of homeostasis, forcing the cell to remodel in order to adapt, given that it is self-referential and self-organizing.

Another way to think of the above is that the cell functions based on a Schrodinger Wave, which is the homolog of the waves generated by the process of homeostasis, fluctuating up and down like a Sine Wave, in order to remain at equipoise with its physiologic set-point.

Human evolution due to bipedalism

Conventional evolution theory dictates that bipedalism was the result of our arboreal ancestors standing erect on tree limbs. But that is reasoning after the fact, which we know is illogical. If the working hypothesis that life has evolved due to communication is correct, then bipedalism ought to connect with that premise, but how?

Turning to the most important transition in vertebrate evolution, the transition from water to land due to the 'greenhouse effect' produced by the accumulation of carbon dioxide in the atmosphere was mediated by three known hormone receptor gene duplications—the Parathyroid Hormone-related Protein (PTHrP) Receptor; the Glucocorticoid (GC) Receptor; the ßAdrenergic Receptor (ßAR). Of these, the most likely to have duplicated first is the PTHrP Receptor, given that the skeletons of boney fish emerging from water onto land was the first trait to experience the transition, given the relative increase in gravitational force from buoyancy to terrestriality. The fossil evidence for that transition is the formation of legs from fins. Over and above the issue of weight-bearing, PTHrP signal amplification was also necessary for land-adaptive traits. PTHrP-deficient mice have skeletal abnormalities both craniofacially and in their limbs, the evolution of the lung alveolus from the swim bladder for air breathing, the evolution of the glomerulus from the glomus of the kidney for micturition, and the evolution of the middle ear bones for hearing in air. The functional relevance for all of these evolved traits is documented by their ontogeny, phylogeny and the causal nature of these traits is exemplified by the deletion of each and every one of these hormone receptors in mice.

PTHrP-deficient mice have skeletal abnormalities both craniofacially and in their limbs. There are a diminished number of proliferating chondrocytes in the growth plate cartilage, as well as accelerated chondrocytic differentiation. To determine the effect of PTHrP deficiency on craniofacial morphology and establish the differential features of the constituent cartilage, the various cartilages in the craniofacial region of neonatal PTHrP-deficient mice were examined. The major portion of the cartilaginous anterior cranial base appeared normal in the homozygous PTHrP-deficient mice. On the other hand, chondrocyte differentiation and endochondral bone formation were observed in the posterior aspect of the anterior cranial base and in the cranial base synchondroses. Ectopic bone formation was found in the soft tissue-running mid-portion of Meckel's cartilage, where the cartilage degenerates and converts to ligament in the course of normal development. The zonal structure of the mandibular condylar cartilage was minimally affected, but the whole condyle was reduced in size. These results reveal that the effect of PTHrP deficiency varies widely among the craniofacial cartilages affected due to the differential features of each cartilage.

The most dramatic difference observed in the Meckel's cartilage of PTHrP-deficient mice was in this soft tissue mid-region. During normal development, the degeneration of this region has been reported to start at around day 18 of gestation in mice lacking type X collagen. Histologic study of the extracellular matrix in rats lacking type X collagen showed that this process of degeneration starts in the perichondrium, where macrophage- and fibroblast-like cells appear to degrade the unmineralized cartilage matrix, and the chondrocytes are finally attacked by giant cells. In homozygous PTHrP-deficient mice, the cartilage in this area was degenerated, and was surrounded by the presumptive bone matrix. Similar ectopic bone formation has also been reported in the perichondrium of rib cartilage in the homozygous mice. These observations suggest that PTHrP deficiency might alter the mechanism of normal bone cell differentiation.

The presence of three ossicles in the middle ear is one of the definitively evolved features of mammals. All reptiles and birds have only one middle ear ossicle, the stapes or columella. How the two additional ossicles appeared in the middle ear of mammals has been studied for the last two centuries, representing one of the classic examples of how structures can change during evolution to function in novel ways. From the combined evidence of the fossil record, comparative anatomy, and developmental biology it is now apparent that the two newly acquired bones in the mammalian middle ear, the malleus and incus, are homologous to the quadrate and articular, which form the articulation for the upper and lower jaws in non-mammalian jawed vertebrates. Incorporation of the primary jaw joint into the mammalian middle ear was only possible due to the evolution of a new way of articulating the upper and lower jaws, with the formation of the dentary-squamosal or temporo-mandibular joint (TMJ) in humans. The evolution of the three-ossicled middle ear in mammals is thus intimately connected to the evolution of a novel jaw joint, the two structures evolving together to form the distinctive mammalian skull.

The middle ear ossicles of mammals reside in an air-filled cavity, straddling the gap between the external and inner ear. Vibration of the tympanic membrane (eardrum) is picked up by the manubrium of the malleus, and transferred to the incus and stapes, conducting the vibrations to the inner ear via the oval window. Defects lead to conductive hearing loss. In birds and reptiles, only one ossicle bridges the air-filled middle ear cavity, passing vibrations from the external to the inner ear. In birds, this ossicle is known as the columella auris, while in reptiles it is known as the stapes.

The middle ear ossicles are found in the auditory bulla, comprised of several bones, namely the tympanic ring, the bulla, and the malleus; the gonial bone lies between the tympanic ring and malleus, facilitating the function of the latter. The malleus derives from both endochondral and gonial bone sources.

Since both reptiles and birds possess only one middle ear ossicle, the origins of the malleus, incus, tympanic ring, and gonial have been controversial. Karl Reichert (in 1837) was the first to hypothesize that the malleus and incus were homologous to the articular and quadrate bones of the non-mammalian jaw joint, which has been supported by extensive interdisciplinary evidence from the fossil record, developmental biology, and molecular biology. Such studies have generated an integrated theory for the mechanisms involved in forming the mammalian ear and jaw.

Evidence from developmental biology

Meckel's cartilage is composed of two rods of cartilage that overarch the sides of the mandible; the proximal portion forms the jaw bone in all but mammalian vertebrates. In avian embryos, portions of Meckel's and the quadrate cartilage derive from the first pharyngeal arch, whereas the retroarticular process that develops proximal to the articular and the columella derive from the second pharyngeal arch. Separated by the jaw joint, cartilage generates the two skeletal elements, giving rise to the quadrate and articular bones of the jaw.

The malleus and incus are formed from a single cartilage that subdivides, whereas the stapes derives from a separate cartilage that extends toward the incus to form a joint. The malleus and incus derive from the posterior of Meckel's cartilage like the other two ear ossicles; the malleus remains attached to Meckel's cartilage throughout most of embryonic development, forming a conduit between the jaw bone and the middle ear.

In mice, the cartilaginous connection between the jaw and ear breaks down after birth, starting on or about day 2 of life, with the transformation of Meckel's cartilage next to the malleus into the sphenomandibular ligament. The dissolution of Meckel's cartilage functionally separates the ear from the jaw. Meckel's cartilage supports the bones that ossify along its length.

Genetic data are also consistent with Reichert's theory. The homeodomain transcription factor Bapx1 is found in the jaw joints of birds, fish, and reptiles, whereas it localizes to the middle ear in developing mammals, pointing to the common origin and homology of the ear ossicles.

The homologies between the ear ossicles suggested by comparative anatomy nearly two centuries ago have been confirmed by molecular and developmental biology. Additional fossil evidence has facilitated further documentation of the transition from the fish jaw to the mammalian middle ear bones.

FoxP2 and the Area of Broca

This holistic perspective on evolution of bipedalism is facilitated by the fact that both locomotion and language formation are determined by the FoxP2 gene. These neurologically controlled traits are localized to the Area of Broca on the underside of the cerebrum. Only great apes have an Area of Broca, and only humans are bipedal, hypothetically facilitated by positive selection pressure for the FoxP2 gene, perhaps initially acting to establish communication on land, followed by its role in locomotion, leading to bipedalism.

Cytoplasm as the 'reference' for Cosmic Consciousness

Assuming that the cytoplasm of the cell is the reference for Cosmic Consciousness raises the possibility that during sleep or meditation the reduction in body temperature facilitates that connection by slowing cytoplasmic streaming. Similarly, dehydration increasing the salt concentration of serum would cause leaching of salt from the cytoplasm, also slowing cytoplasmic streaming. Those processes would hypothetically enhance the shift from local to non-local consciousness. Of course it must be kept in mind that such physiologic changes may also cause hormonal changes, particularly the posterior pituitary hormones oxytocin and vasopressin, raising the possibility that they further facilitate such transcendent conscious experiences.

Chapter 17
Implicate and Explicate Orders as Unconscious and Conscious

Introduction

Science asks us to suspend our commonly held beliefs in favor of falsifiable evidence (Popper, 1959) for good reason. We began life as an ambiguity of energetic states as negative entropy (Torday and Miller, 2017), and have been making up 'Just So Stories' to cope with that reality ever since.

But there is an ultimate Truth contained within the Cosmos that physicist David Bohm describes as the Implicate Order (Bohm, 1980). We tend to universally express that concept as 'something greater than ourselves', but the answer to this enigma has thus far remained intractable. However, scientific breakthroughs based on empiricism continue to encourage us to seek that ultimate Truth, such as Heliocentrism (Kuhn, 1957), The Periodic Table of Elements (Scerri, 2019), Evolutionary Biology (Darwin, 1859), Quantum Mechanics (Mehra and Rechenberg, 1982) and The Big Bang (Dodelson and Schmidt, 2020). And if there were some common origin from which all of them were to emanate, we should theoretically be able to identify it by finding self-similar patterns in each of them. That commonality is tentatively identified herein as the patterns formed by both the Periodic Table of Elements and Evolutionary Biology, originating from the way that gravity enables Quantum Entanglement (Torday, 2021)—when gravity strikes the curved surface of a protocell it produces energy based on Einstein's Field Theory (1961). That energy sustains the Quantum Entanglement for the contents of the cell, in turn referencing the non-local gravitational force of the Cosmos.

Such hope for understanding life's mystery has been buoyed up by the recent realization that the cell membrane behaves on two levels—as a barrier for the entry and exit of matter in and out of the cell, and as a 'mobius strip' that accommodates the holistic nature of the cell as the conscience of the Cosmos (Torday, 2021). In that vein, it has been hypothesized that our physiology constitutes Bohm's Implicate Order, whereas our day-to-day behavior constitutes Bohm's Explicate Order. That way of understanding our state of being would explain the unconscious and conscious, for example, as the Implicate and Explicate Orders, respectively. And how and why under stress conditions the two states merge together to one degree or another due to the influence of the endocrine system acting to mediate wave collapses due to the merging of the calcium fluxes within cells, neurons and non-neurons alike (Torday, 2022). Such events as Near-Death Experiences, Out-of-Body Experiences, Maslow Peak Experiences, and the Runner's High may all be consequences of hormonally-mediated wave collapses. Beyond that, the two-tiered dual nature of consciousness may explain the Double Slit Experiment, the particle beam of light complying with the Explicate Order, the wave form with the Implicate Order. And the Libet Experiment (Libet et al., 1983) may be interpreted as experimental evidence for such a hypothetical duality of consciousness. The 300 millisecond time lag between sensing the electrical shock and the response may be due to our Implicate consciousness response.

These insights have been revealed by an empiric approach to the evolution of physiology based on the principle of cell-cell communication, mediated by soluble growth factors and their receptors (Torday and Rehan, 2012; Torday and Rehan, 2017). As such, this approach is unique among the many ways in which evolutionary biology has been addressed since it is solely based on empiric evidence, and is therefore faithful to Popperian testability and refutability (Popper, 1959).

The Periodic Table of Elements and You

Eric Scerri has written extensively about Mendeleev's Periodic Table of Elements. In his book "The Periodic Table: its story and its significance", Scerri (2019) importantly informs us that Mendeleev did not merely arrange the Elements based on their atomic number; he also used how they behaved chemically to produce specific salts to guide his decisions in constructing his version of the Periodic Table, taking into account variations like isotopic forms, reinforcing the diachronic across-spacetime use of atomic number to arrange the Table. Atomic Number is a function of the number of protons in the nucleus of each Element, which references its diachronic origin.

Nucleosynthesis is the means by which stars produce light, in the process forming the Elements, arranged from the lightest to the heaviest (Smolin, 1999), from hydrogen to iron. That progression provides a 'logic' for the Cosmos, and the process of Symbiogenesis (Margulis, 1993) provides the means by which organisms have coped with existential threats by endogenizing and compartmentalizing them to form physiology. Husserl tells us that this is the nature of geometry, for example, which exists within our physiology, but is manifested outward as lines and circles and triangles, not unlike Plato's archetypes. Such conscious properties are so ancient that we do not recognize their origins any longer. Such physiologic 'elements' must conform to the same physical logic by obeying the Laws of Nature.

Such use of empirical knowledge is akin to the way that evolutionary biology has been reverse-engineered based on cell-cell communication mechanisms in developmental biology (Torday and Rehan, 2012; Torday and Rehan, 2017), the three primary germ layers—endoderm, ectoderm and mesoderm-interacting with one another sequentially, beginning with the fertilized egg or zygote, ultimately giving rise to the offspring. The premise is that embryology is the only mechanism we know of for generating form and function biologically. By superimposing the developmental cell-cell signaling mechanisms that determine form and function onto the phylogenetic changes in phenotype as speciation using the same signaling mechanisms as a common denominator (Torday and Rehan, 2007), the underlying principles of change from the swim bladder to the lung, or the glomus of the fish kidney to the glomerulus of land animals emerge. And like the empiric details of the Periodic Table mentioned above, the hormonal effects on evolution provide the dynamic aspects of phenotypic change both ontogenetically and phylogenetically, independent of the time differences (Torday, 2022).

The inherent interrelationships between physics, chemistry and biology are illustrated in Figure 1. Subsequent to the Big Bang, depicted at the bottom of the figure, the Elements were formed through nucleosynthesis (Smolin, 1999), and distributed throughout the Cosmos. Organisms were able to cope with the environment by endogenizing physical factors that posed an existential threat. The Elements in the Periodic Table represent both their "synchronic" physical characteristics, and their "diachronic" number of protons in their nuclei.

Similarly, evolution is constituted synchronically as embryologic development, and diachronically as phylogeny. In this sense, the Periodic Table of Elements and the process of evolution are homologs, being derived from the same origin. This is consistent with the homology between the atom and the cell (Torday, 2018a), which are both point sources that adhere to Pauli Exclusion Principle and Heisenberg Uncertainty Principle, conferring deterministic and probabilistic characteristics to each (Torday and Miller, 2016b).

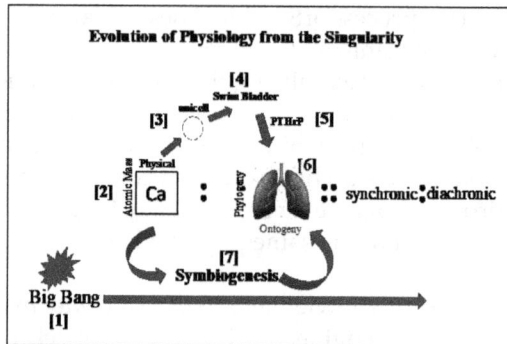

Figure 1. Origin of Physical, Chemical and Biological Homologies. Beginning with the Big Bang [1], the Elements formed through nucleosynthesis [2] synchronically and diachronically as both physical characteristics and Atomic Mass. Unicells [3] gave rise to complicated physiology, such as the swim bladder [4], PTHrP [5] facilitating evolution into the lung [6] mediated by symbiogenesis [7]. The Elements and biology are both synchronic and diachronic, indicating their homology [from 2 to 6].

Parathyroid Hormone-related Protein (PTHrP) epitomizes the cell-cell interactions that have mediated the evolution of vertebrate structure and function, as represented by the morphing of the fish swim bladder into the mammalian lung, the first 50 genes involved in the development of each being identical (Zheng et al., 2011).

These are conventionally thought of as time-based processes, but since Einstein tells us that time does not exist (Canales, 2016), these properties of biology exclusively reference space (Rowlands, 2015).

The question as to whether time exists in physics and biology or not has been discussed extensively. Einstein and Bergson debated this question publicly in 1922 (Canales, 2016). Because of the way that epigenetic inheritance works in service to the maintenance of the unicellular state (Torday, 2015a), its emerging prominence would favor the non-existence of time in biology as an artifact of descriptive biology rather than mechanistic biology. The irrelevance of time in biology is consistent with the counterintuitive idea that the future determines the present (Di Corpo and Vannini, 2015). Our book (Guex et al., 2020) on "Morphogenesis, Environmental Stress and Reverse Evolution" provides empiric evidence for 'reverse evolution', challenging the existence of chronological time in biology.

It is interesting to recall here the speculation of William Crookes in his 1886 Presidential Address to the Chemical Section of the British Association that a spiral representation of the Periodic Table could be explained in terms of a progressive evolutionary genesis of the elements as a result of two forces, 'operating in accordance with a continuous fall of temperature', and the other showing a sinusoidal variation (simple harmonic oscillator), connected with the electric force, together producing a (double helical) generation of elements of increasing atomic mass, but periodically similar chemical properties (Scerri, 2019). This speculation, which predated the discovery of basic atomic structure—the electron, the proton, and neutron—which predicted isotopes and accommodated the (undiscovered) inert gases, could now be seen as exemplifying the characteristic actions of Nature when operating according to the (then unknown) universal 'rewrite system' chronicled by Peter Rowlands in his book on The Foundations of Physical Law (2015).

Thinking about how the dual forces of decreasing entropy and the sinusoidal oscillations/electrical force yielding a double helical generation of elements would coincide with the biology, such oscillations are reminiscent of the oscillating levels of oxygen in the atmosphere over the last 500 million years, fluctuating between 15 and 35% (Berner, 1999). The documented periodic increases in oxygen caused the widely recognized phenomenon of 'giantism' (Vermeij, 2016), whereas the physiologic effects of the periodic decreases in oxygen are not addressed anywhere in the scientific literature, other than what has been hypothesized regarding the evolution of endothermy (Torday, 2015). Hypoxia is the most powerful natural physiologic stressor known. That correlates with specific physiologic changes that occurred in the Pituitary-Adrenal Axis during this same epoch—the appearance of the Parathyroid Hormone-related Protein (PTHrP) gene in both the anterior pituitary (Mamillapalli and Wysolmerski, 2010) and adrenal cortex (Mazzocchi et al., 2001). The structural-functional effect of PTHrP is seen in the capillary arcades of the adrenal medulla (Wurtman, 2002), which expanded sometime during this era as well. That is physiologically relevant because the hormonal secretions of the adrenal cortex pass through the vascular arcades of the adrenal medulla on their way out of the adrenal gland to the systemic circulation. Since PTHrP stimulates the formation of capillaries (Diamond et al., 2006), the PTHrP produced in the adrenal cortex would consequently have increased the vasculature of the medulla. As a result of the above, the effect of the corticoids produced in the adrenal cortex passing through the adrenal medulla, stimulating the rate-limiting step in adrenalin production (Sharara-Chami et al., 2010), increasing adrenalin secretion by the medulla, would have been amplified by the increased surface area of the adrenal medullary vasculature (see above).

Adrenalin stimulates the production of lung surfactant by the alveoli (Lawson et al., 1978), increasing their distensibility, and consequently their rate of oxygen uptake. So, in the aggregate, this cascade is primarily in service to alleviating the episodic hypoxia caused by the step-wise evolution of the nascent lung in adaptation to land in the short-run.

However, in the long-run, PTHrP increases the formation of alveoli (Rubin et al., 2004), constitutively increasing oxygenation.

As for why that all may have occurred, in hindsight we evolved from small, dog-sized cynodonts that had to be nimble and quick to survive, hence the adaptive amplification of the fight or flight mechanism.

The environment gave rise to endothermy

All of the above has been incorporated into a "Central Theory of Biology" for the evolution of warm-bloodedness, or endothermy/homeothermy (Torday, 2015b). Briefly [see Figure 2], the adrenalin that alleviated the lung alveolar constraint on oxygenation also increased the release of fatty acids from fat stores (Carey, 1998). Fatty acids are the optimal substrate for metabolic production of heat, so the increased metabolic activity would have raised body temperature, ultimately evolving genetic control as the thermoregulatory action of oxytocin (Kasahara et al., 2013), a neuroendocrine hormone produced by the posterior pituitary gland (Leng et al., 2015).

The ability to maintain body temperature independently of the environment fostered the transition from cold- to warm-bloodedness due to more efficient metabolism since multiple enzyme isoforms are needed for any given metabolic step in cold-blooded organisms in order to optimize the metabolic activity at different ambient environmental temperatures, whereas warm-blooded organisms only require one form of any given metabolic enzyme.

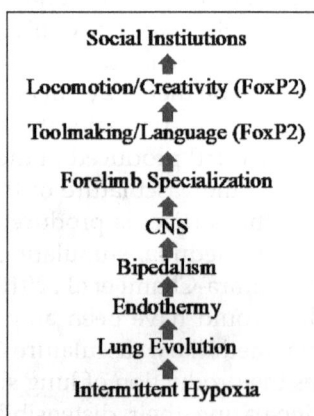

Social Institutions
⬆
Locomotion/Creativity (FoxP2)
⬆
Toolmaking/Language (FoxP2)
⬆
Forelimb Specialization
⬆
CNS
⬆
Bipedalism
⬆
Endothermy
⬆
Lung Evolution
⬆
Intermittent Hypoxia

Figure 2. Central Theory of Biology. Starting with intermittent hypoxia, the evolution of land vertebrates advanced due to endothermy, allowing for bipedalism, freeing the forelimbs for toolmaking, including language and imagination codified as the gene FoxP2. The development of written language facilitated the rise of human social institutions, allowing for the transfer of knowledge from generation to generation.

Increased metabolic efficiency facilitated bipedalism because it takes more energy to walk on two legs than on four (Rodman and McHenry, 1980), resulting in the freeing of the forelimbs for specialized adaptations such as flight in the case of birds, and toolmaking in the case of humans. That placed additional selection pressure on the central nervous system to integrate and control such complicated physiologic properties. The upshot of that selection pressure was language as a form of toolmaking, particularly the written word, which codified the myths we use to cope with our ambiguous origin referred to above. And because such hard copies of our narratives can be passed from generation to generation, they become part and parcel of our institutionalized belief systems.

Only great apes have a structure on the underside of the cerebrum called the Area of Broca that controls both motor control and language (Palomero-Gallagher and Zilles, 2019). In humans, the coalescing of toolmaking, locomotion and language is under the common control of FoxP2 (Xu et al., 2018), underscoring the positive selection for this constellation of traits in humans. And beyond that, FoxP2 gives rise to human creativity (Lieberman, 2009). It appears as though the combination of language and bipedal locomotion fosters our curiosity, giving rise to imagination.

The progressive increase in entropy in the environment due to the Second Law of Thermodynamics placed ever-greater selection pressure on biology to amplify the First Principles of Physiology (Torday and Rehan, 2009) that initially permitted negative entropy within the cell. That, in turn, would have been dependent on the lipid facilitation of oxygenation, initiated by the insertion of cholesterol into the cell membrane (Bastiannse et al., 1997), followed by the evolution of peroxisome-derived lipids to protect against the rising oxygen within the cell (De Duve, 1969), lipids as substrate for steroid hormones (Hume and Boyd, 1978), the endocrine system (Kalra and Priya, 2018), and physiologic evolution (Torday and Rehan, 2012; Torday and Rehan, 2017).

The recognition of lung surfactant evolution as a series of pre-adaptations (Torday and Rehan, 2009) provided deep insights to the fundamental interrelationship between lipids and oxygen uptake, from the composition of lung surfactant lipids (Orgeig et al., 2015) all the way back to the unicellular state based on the biosynthesis of cholesterol (Torday, 2013), the most primitive of lung surfactants (Orgeig and Daniels, 2001); Konrad Bloch hypothesized that cholesterol was a 'molecular fossil' since it takes 11 atoms of oxygen to produce one molecule of cholesterol (Bloch, 1992). Therefore, there had to have been enough oxygen in the atmosphere to do so, linking oxygen and cholesterol together mechanistically in space-time.

Overall, that process, in turn, gave insight to the evolution of many other physiologic traits, particularly those facilitated by Parathyroid

Hormone-related Protein (PTHrP), namely the lung, kidney, skin, brain and skeleton (Guise and Mundy, 1996). Such insights led to a focus for vertebrate evolution on the water-land transition (Romer, 1949), during which the PTHrP Receptor gene duplicated (Pinheiro et al., 2012), or essentially amplified due to its property as a receptor.

Tiktaalik provided scientific evidence for the fossilized remains of the transition from fish to tetrapods (Daeshler et al., 2006), but there are no fossil data for the modifications of the internal organs that occurred during that transition in adaptation to land since such soft tissue structures would not have been preserved. However, there are extensive data for the cellular-molecular development of the organs that were essential for land adaptation, such as the lung and kidney. When such developmental mechanisms are superimposed on phylogenetic changes, they reflect the underlying cellular-molecular changes that occurred over the course of evolution (Torday and Rehan, 2007). Such hypotheses have been corroborated by gene deletions and over-expressions that are consistent with such evolutionary changes.

The Periodic Table and evolutionary biology are diachronic vectors of the Big Bang

The use of PTHrP signaling diachronically across space-time is homologous (of the same origin) with Mendeleev's use of chemical reactions to construct his Periodic Table of Elements. More importantly, receptor signaling through such 'second messengers' as cyclic Adenosine Monophosphate and inositol phosphates (Rubin et al., 1994) gave even deeper insight to other such cell-cell signaling mechanisms occurring in tandem in other tissues and organs (Torday and Rehan, 2012). The magnitude and direction of these signaling pathways are vectors for biologic change, referring all the way back to the unicellular state (Torday and Miller, 2016), which dominated life on Earth for the first 3.5 billion years (Mojzsis et al., 1996). That pattern is comparable to the way in which chemical reactions aided Mendeleev's assembly of the Periodic Table of Elements. Importantly, both the chemical and biologic reactions are diachronic mechanisms that cut across space-time. The vectoral magnitude and direction of those chemical and biologic reactions ultimately reference the Singularity (Torday, 2018a), from which they emerged as asymmetries produced by the Big Bang. Admittedly, there is a 10 billion year gap between the Big Bang and the origin of life, but the case has been made for the homology between the atom and the cell (Torday and Miller, 2016b), providing the structural-functional bridge for that gap.

For example, this may be why William of Ockham declared that the simplest answer is probably correct (Spade, 1999). Those observations that are most consistent with the vector of the Big Bang represent the shortest distance between two points, namely the Singularity and physical, chemical, and biologic reactions.

Mathematical expression of such vectors would lead to a way of 'calculating' the value of any given property of Nature. Peter Rowlands, a physicist at the University of Liverpool has provided just such a mathematical system (2015), arguing that there are fundamental principles of physics that underlie all we experience. In his reduction, all of reality can be expressed in terms of 'zeros and ones'. We conventionally focus on the 'ones' as our material reality, but it is actually zero, which Rowlands refers to as an attractor, or organizer, that is the key to understanding the fundament of physics. That is because material things are merely epiphenomena of the energy produced by the Big Bang, matter being a by-product due to Newton's Third Law of Motion- every action having an equal and opposite reaction. Absent that 'recoil', the Cosmos would only be composed of energy. The same holds true for biology; witness the morphologic changes that occur during embryology, beginning with the fertilization of the egg, characterized by a huge flash of energy due to the 'zinc spark' (Duncan et al., 2016). The subsequent cell-cell interactions that mediate embryologic growth and differentiation are mediated by growth factors triggering high energy phosphates. Focusing on the flow of energy exchanges rather than on the cellular changes provides a much simpler perspective on life (Torday, 2016) than Darwin's 'tangled bank', particularly when it is considered that the developmental cell-cell interactions culminate in physiologic homeostasis as the energetic balance for the rest of the life cycle, the organism serving to collect epigenetic data from its environment (Torday and Miller, 2016a).

When thought of in those terms, focusing on the cell as the basis for biologic evolution amounts to the same thing because the negative entropy within the cell is 'zero' relative to the positive entropy outside of the cell (Schrodinger, 1944). Negative numbers are a mathematical construct that is not relevant to biology. Therefore, the negative entropy within the cell is virtually zero, but no less. Therefore, both the animate and inanimate can be reduced to the same set of fundamental principles. Indeed, this way of thinking about Cosmology is synonymous with Alfred North Whitehead's Process Philosophy (Whitehead, 1929). He thought that everything in the Cosmos was energy, and that matter occasionally appears coincidentally, but is transient.

Whitehead thought that our focus on the material at the expense of the non-material was misguided because it did not engender relationships. In his book "Science and the Modern World", Whitehead (1925) expressed the idea that matter is 'senseless, valueless, purposeless' because it is merely a transient by-product of the underlying energetic forces of the Cosmos. He called this perspective 'scientific materialism'. Whitehead found fault with the irreducible nature of matter because it masks the importance of change since nothing ever stays the same. Whitehead placed the emphasis of reality on change, and that 'all things flow'. This concept was first voiced by Heraclitus, who said that "No man ever steps in the same river twice, for it is not the same river and he is not the same man" (Kahn, 1979).

Whitehead thought that viewing objects as separate and distinct from all other objects was a systematic error because each object is an inert mass that is only superficially related to other things. The idea that the material is the primary state of being leads people to conclude that objects are all separated by time and space, and are not related to anything.

Conversely, Whitehead thought that relationships were the primary state of being. Whitehead describes any entity being nothing more or less than the aggregate of its relationship to other entities, as the synthesis of and reaction to the world around it. Relationships are not secondary to what a thing is, they *are* what the thing is. Robert Rosen held a similarly strong opinion about relationships (Rosen, 1958).

Whitehead's concept of reality is consistent with Mendeleev's way of assembling the Periodic Table (Scerri, 2019). He saw Elements in the context of their reactivity with other elements rather than as inert matter, with certain superficial characteristics like those described by Alchemists. In that sense, it was the Alchemist who determined how to make gold out of dross, not the innate characteristics of the Elements that enabled them to interrelate with one another.

Similarly, by viewing organisms through the lens of their abilities to recapitulate themselves developmentally, the processes by which they fundamentally interrelate with one another comes into view. That is particularly true when development is seen in the larger context of phylogeny, offering the retrograde retracing of the process of evolution back to the unicellular state as a series of pre-adaptations (Torday and Rehan, 2012). Such interactions between the organism and its environment over the course of evolution subordinate the material in favor of the process.

Therefore, now it can be stated unequivocally that both the Periodic Table of Elements and Evolutionary Biology are expressions of the process of change rather than descriptions of the superficial appearance of material objects- elements and phenotypes. Whitehead was thus correct in his perspective, but he had no experimental evidence to scientifically substantiate what he was espousing. Now, armed with evidence for this comprehensive view of reality, it behooves us to practice Whitehead Process Philosophy.

Information Theory Meets Informatics

Information theory is the epitome of Whitehead's focus on process since it represents the mechanism for forming, storing and sharing information (Shannon, 1948). However, because it subordinates the transmission of the information to the materialism of the information it is misguided. The consequence of that is the discipline of Informatics (Arndt, 2004), which tends to be conflated with knowledge. This is syllogistic reasoning that has undermined human thought and action. The prevailing concept in Informatics is that if you have not solved the problem you just need more data. That may be true at NASA, where Informatics was developed to keep track of the parts for assembling the space shuttle, but does not apply to biologic systems. In the case of the shuttle, the data are a closed set, whereas biology is an open set, the whole being greater than the sum of its parts.

For example, hormones do not carry information, they merely facilitate communication between cells, triggering the energy bursts of 'second messengers', which are high energy phosphate compounds that bind to DNA elements in the nucleus and affect embryologic growth and differentiation. It is only once the hormone binds to its receptor that information within the cell is made available. Therefore, the information lies within the target cell itself, not in the hormone.

Similarly, in the article about "Phenotype as Agent" (Torday and Miller, 2016a), it is argued that we misconstrue what phenotype is based on the description rather than the mechanism of evolution. Phenotype is the agency of the Explicate Order for detecting environmental change, whereas convention says that it's the material composition of the organism. In this frame, perhaps our physiology is the product of the Implicate order, acting as the reference point for the phenotypic Explicate Order. The residue of the Implicate Order remains in our physiology in various forms, such as the mesoderm, which is introduced between the endoderm and ectoderm as the third germ layer during gastrulation. Lewis Wolpert said it was 'truly the most important time in your life', because the mesoderm confers the plasticity to adapt. And then there's the advent of deuterostomes-developing from posterior to anterior, in the opposite direction to the force of gravity. That is homologous with the effect of bipedalism, freeing the forelimbs to specialize, again placing positive selection pressure anteriorly for further evolution of the head and brain (Torday, 2015b). That process is mirrored and reinforced by the vagus, the major nerve of the Autonomic Nervous System, which emanates from the adrenals, to the gut, to the heart, terminating in the head (Porges, 1995).

Truth be told

It is because of the relationship between materialism and process that we have been able to advance as a species among species. However, as David Bohm points out, we continue to exist within the Explicate Order, made subjective by our evolved senses, whereas there is a true reality just beyond our reach, referred to as the Implicate Order (Bohm, 1980). It is because of the deceptive reality we have formulated (Trivers, 2011) that we must periodically rise up, and then fall, ultimately succumbing to the Laws of Nature. Whether it is Climate Change or economics, we are vulnerable to our failure to comply fully with the prevailing forces of Nature. We come ever-closer to those Laws as we evolve and endogenize the environment over the course of evolution, but in the current environment of the Anthropocene (Zalasiewicz et al., 2019) our narcissistic tendencies are out-stripping our natural arc for lack of insight regarding process over materialism (Whitehead, 1929). Prior to the use of fossil fuel to drive our machines, there was an intimate, demonstrable interrelationship between the energy of the earth and that of life (Torday and Rehan, 2011). Subsequently, the introduction of man-made thermal energy into the environment caused a dissociation of Man from our evolutionary trajectory. Consequently, once we begin engineering our heredity using CRISPR (Zhang et al., 2014) we will deviate from our naturally evolved trajectory from the Singularity, perhaps evolving as 'silicon-based life forms' instead.

There is only space, There is no time

If, as Rowlands has concluded, the only directly measurable dimension is space (Rowlands, 2015), Einstein and Feynman having cast doubt on the existence of time, how do we explain the time-based understanding of development and phylogeny? It would have to be assumed that the latter are exclusively space-filling properties of biology. In this vein, we have learned that epigenetic inheritance directly from the environment affects evolution, and it has been proposed that it fosters 'running in place' to maintain homeostasis in consilience with the First Principles of Physiology (Torday and Rehan, 2009), which would subsume a spatially non-temporal way of thinking about biology. This question is reminiscent of the well-documented debate between Einstein and Bergson in 1922, Einstein insisting that time is an artifact of biology, Bergson countering that time is critical for understanding any and all of biology and psychology (Canales, 2016).

If, as has been proposed (Torday, 2019), The Singularity is the prototype for biology, and evolution is the process for remaining faithful to it, striving to emulate the Singularity, then time would drop out of the 'equation', leaving space as a point source at its minimum (Torday and Miller, 2016b), the Cosmos as its maximum.

A novel prediction of consciousness as the singularity

The premise for the idea that Consciousness is the expression of the Singularity is that there is an intersection of the Singularity with physiology, the latter as the effective endogenization of the cosmologic environment (Torday, 2020), at least those features that have posed existential threats to the organism. That is to say, when the Singularity was disrupted by the Big Bang [see Figure 3], the information therein was fragmented, but still ultimately had to conform with the Laws of Nature. When life began on earth some 4.5 billion years ago, it too had to conform with the Laws of Nature. It did so by endogenizing the environment and making it useful by compartmentalizing it as what we refer to as physiology (Margulis, 1993). In the aggregate, our physiology ascribes to the Singularity as its origin (Torday, 2019), and the way in which physiology functions to maintain homeostasis is based on the same set of principles (Torday, 2015c). In other words, our interoceptive sense of self is founded on the Singularity as the origin of consciousness, actualized by the physiologic principles that have evolved from the Cosmos (Torday, 2018b). In turn, our agency as conscious beings gains purchase to on-going variations in our ever-changing environment.

Holistically, our physiology maps onto the Singularity since it is the product of endogenized physical factors in the environment that have

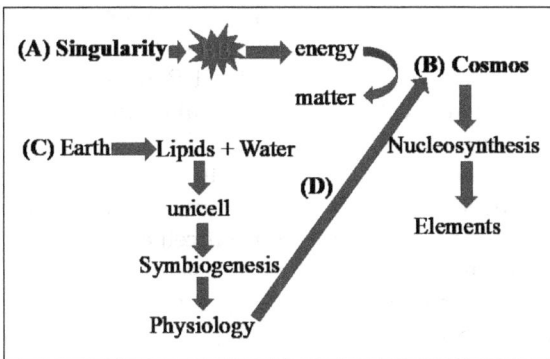

Figure 3. The Origin of Consciousness. The Big Bang (A) gave rise to energy and matter. (B) Nucleosynthesis, the formation of light by stars, generated the Element is sequence from lightest to heaviest, providing a 'logic'. (C) Cells were formed from lipids immersed in water; when posed with existential threats, cells assimilated them through Symbiogenesis. Such factors formed physiologic systems, (D) but must comply with the same Laws of Nature as the Cosmos, forming the connection between life and non-life.

been compartmentalized and made functionally useful (Torday, 2019). Physiology functions to maintain homeostasis based on the same set of natural laws that the atom ascribes to (Torday, 2015c). When lipids spontaneously formed micelles, distinguishing the inside of the cell from the outside environment, fueled by negative entropy (Schrodinger, 1944), life began, initiating the Explicate Order in the process (Bohm, 1980). Consequently, our interoceptive sense of 'inside' connects to the Singularity as consciousness, actualized by the physiologic principles that have evolved from the Cosmos (Torday, 2018).

It is the principle of homeostasis that integrates all of these properties—when the Big Bang occurred some 13.8 billion years ago, there was an 'equal and opposite reaction' to it based on Newton's Third Law of Motion. That reaction is the pre-adaptation that we refer to as homeostasis, acting to entrain inanimate balanced physical, chemical, and biologic reactions alike (Torday, 2015c). Without homeostasis there would be no matter, there would only be energy. Alfred North Whitehead's 'Process Philosophy' states that all is energy; matter is a transition between energy states (Whitehead, 1929).

The vertical integration of gravity, chemistry and biology as consciousness

As mentioned above, PTHrP integrated numerous physiologic properties in vertebrates due to evolutionary selection pressure during the water-land transition. This is attributable to its mechanotransduced effects on the skeleton and lung. Such vertically-integrated physiologic mechanisms offer deep insights to the fundamental interrelationships between physics, chemistry and biology. For example, when lung or bone cells are exposed to microgravity, PTHrP messenger RNA decreases rapidly, interrupting the cell-cell communication mechanisms it mediates for breathing (lung), salt/water balance (kidney), and bone calcification, for example (Torday, 2003). In addition to being a mechanotransducer for cell-cell interactions, PTHrP is a calcium regulatory paracrine factor, acting synergistically with its effect on cell-cell communication to ensure that calcium levels in the alveolar hypophase are appropriate for optimal lung surfactant surface tension lowering function, and for calcification of the skeleton in response to PTHrP signaling in bone (Torday, 2003). Even more fundamentally, when yeast are exposed to microgravity they lose their ability to generate a calcium flux or reproduce, leaving them 'comatose' (Purevdorj-Gage et al., 2006). Molecularly, mechanical forces like gravity affect the cell via the Target of Rapamycin (TOR) gene, which interacts with the cytoskeleton (Diz-Muñoz et al., 2016). In turn, matrix proteins in the extracellular space interconnect with cytoplasmic signaling mechanisms, orchestrating the myriad chemical reactions that govern cell physiology on a moment to moment basis. Thus, the gravitational force produced by the Big Bang references the Singularity as the physiologic unity being emulated by the unicell, endogenized over the course of evolution, ultimately generating Consciousness as the holism of being (Torday and Miller, 2018).

For example, it has been shown that gravity is necessary for life since differentiated lung and bone cells lose their phenotypic identity in zero gravity (Torday, 2003). Similarly, Purevdorj-Gage et al. (2006) have shown that yeast, which are simple eukaryotes, exposed to zero gravity cannot conduct a calcium flux because they cannot polarize, and cannot reproduce, rendering them non-functional. Even the purely chemical micelle cannot form normally in the absence of gravity (Claassen and Spooner, 1996). That is because the fundament of Symbiogenesis is Quantum Entanglement, which entrained particles within micelles at the beginning of life (Torday, 2021), fueled by gravity impinging on the curved surface of the micelle (Misner et al., 1973), the particles synchronizing themselves non-locally with events in the Cosmos. This continuum from the origin of life to the Cosmos is where consciousness evolved from (Torday, 2018).

Biology and chemistry as vectoral fractals of the Big Bang

As mentioned at the outset, the genius of Mendeleev's Periodic Table of Elements was not only in his use of atomic weight as the organizing principle, but in further fine-tuning it based on the chemical reaction products that further characterized each element (Scerri, 2019). That empiric approach offered a much more dynamic way of calibrating the Elements diachronically across space-time rather than merely arranging them based on their physical characteristics synchronically.

The same holds true for using the cell-cell interactions that generate form and function developmentally, applied to phylogeny to understand evolution, which is also a diachronic perspective. The value added by this perspective is in being able to eliminate time from the analysis, leaving only the dimension of space to be considered; and space can be reduced to a point-source without any dimensions, as pure energy, consistent with Whitehead's 'Process Theory'. In contrast to that approach, forming cladograms merely describes the physical progression of evolution without any underlying understanding of how or why it occurred at the cellular-molecular level.

Moreover, the use of such empiric data reflects the changes in the chemical and biologic reactants to their products. When shown graphically, those reactions are vectors for the magnitude and direction of change. It is hypothesized that those vectors are fractals of the Ur-vector formed by the Big Bang. And the more proximate the chemical and evolutionary biologic vectors are to the vector formed by the Big

Bang, the more fundamental they are. Yet like Zeno's Paradox, we can never be evolutionarily congruent with the ultimate vector of the Big Bang because we ironically must evolve in response to vectoral changes in the environment or become extinct.

This is why William of Occam's Razor is indicative of the right solution, the shortest distance between two points being the shortest, simplest path.

Terminal Addition and Phantom Limb as 'Proof of Principle' for physiology as our implicate order

The phenomenon of Terminal Addition, or the appearance of a newly acquired trait at the end of a series of evolved traits, is well recognized in biology, yet how and why it occurs has not been explained. It stems from the fact that embryologic development is achieved by cell-cell communications mediated by soluble growth factors signaling to their cognate receptors, stimulating the production of high energy phosphates as 'second messengers' for growth or differentiation. The most efficient way to introduce such a new trait is by adding it on to the end terminus because inserting in the middle or at the beginning would disrupt the predecessors, requiring them to rewire, which is highly inefficient, putting the organism at risk of extinction, whereas adding such new traits to the end is relatively easy. In so doing, it provides a 'logic' founded on the assimilation of environmental factors, aligning the internal environment of the cell with the external environment of the environment, providing a natural homology between the cell and the Cosmos.

A concomitant of Terminal Addition is the phenomenon of "Phantom Limb Syndrome", an amputee having an itch in the big toe of a severed leg. At first glance, this sense seems highly inefficient. However, when seen in light of Terminal Addition, loss of such a phantom sensation would result in loss of sensation in the traits up-stream from the amputated limb too, causing their atrophy. So for the sake of conservation of structure and function, an imagined input from the missing limb is maintained in order to preserve the integrity of the organism in the vein the phenotype as agent.

Conclusions

In an earlier publication it had been suggested that regressing the data for cellular-molecular lung evolution would approximate the origin of life at the intersection of its Cartesian Coordinates (Torday and Rehan, 2007). By reverse-engineering physiologic traits like the lung, tracing the cell-cell communications from the alveolus back to the swim bladder, and further back to the cell membrane of unicellular eukaryotes based on lipids facilitating oxygen uptake [see Figure 1]. Using that approach allows a 'back-calculation' as far back as Quantum Entanglement and non-localization as pre-adaptations for Symbiogenesis.

The present claim for that prediction is also based on cellular evolution, but now as serial pre-adaptations, specifically identifying the Singularity as the ultimate pre-adaptation of the unicellular state (Torday, 2019). As a 'point source' (Torday and Miller, 20176), the Singularity would occupy no space, merely being a locale of energy essentially equaling zero space. Rowlands has similarly suggested that in math, zero is an attractor, acting as an organizing principle (Rowlands, 2015).

Such metaphysical ideas might help in identifying the actual nature of Consciousness, which has remained intractable for thousands of years, beginning with the Ancient Greek Philosophers, right up to today, when philosophers like David Chalmers (Chalmers, 1996) and Andy Clark (Clark, 1998) pose "hard" questions about qualia. In the current context, it is proposed that Consciousness lies at the intersection of Cosmology and Physiology, the product of which is what we think of as being conscious, or mind.

It is offered that chemical and physical properties are vectoral fractals of the Big Bang, which could be tested by mathematically modeling key reactions for each.

Acknowledgements

John S. Torday has been funded by NIH Grant HL055268.

References cited

Arndt, C. 2004. Information Measures, Information and its Description in Science and Engineering. Springer, New York.

Bastiaanse, E.M., Höld, K.M. and Van der Laarse, A. 1997. The effect of membrane cholesterol content on ion transport processes in plasma membranes. Cardiovasc. Res. 33: 272–283.

Berner, R.A. 1999. Atmospheric oxygen over Phanerozoic time. Proc. Natl. Acad. Sci. U. S. A. 96: 10955–10957.

Bloch, K. 1992. Sterol molecule: structure, biosynthesis, and function. Steroids 57: 378–383.

Bohm, D. 1980. Wholeness and the Implicate Order. Routledge, London.

Canales, J. 2016. The Physicist and the Philosopher. Princeton University Press, Princeton.

Carey, G.B. 1998. Mechanisms regulating adipocyte lipolysis. Adv. Exp. Med. Biol. 441: 157–170.

Chalmers, D. 1996. The Conscious Mind. Oxford University Press, Oxford.

Clark, A. and Chalmers, D.J. 1998. The extended mind. Analysis 58: 7–19.

Claassen, D.E. and Spooner, B.S. 1996. Liposome formation in microgravity. Adv. Space Res. 17: 151–160.

Daeschler, E.B., Shubin, N.H. and Jenkins, F.A. Jr. 2006. A Devonian tetrapod-like fish and the evolution of the tetrapod body plan. Nature 440: 757–763.

Darwin, C. 1989. On the Origin of Species. John Murray, London.

De Duve, C. 1969. Evolution of the peroxisome. Ann. N.Y. Acad. Sci. 168: 369–381.

Diamond, A.G., Gonterman, R.M., Anderson, A.L., Menon, K., Offutt, C.D., Weaver, C.H., Philbrick, W.M. and Foley, J. 2006. Parathyroid hormone hormone-related protein and the PTH receptor regulate angiogenesis of the skin. J. Invest. Dermatol. 126: 2127–2134.

Di Corpo, U. and Vannini, A. 2015. Syntropy. ICRL Press, Princeton.

Diz-Muñoz, A., Thurley, K., Chintamen, S., Altschuler, S.J., Wu, L.F., Fletcher, D.A. and Weiner, O.D. 2016. Membrane Tension Acts Through PLD2 and mTORC2 to limit actin network assembly during neutrophil migration. PLoS Biol. 14: e1002474.

Dodelson, S. and Schmidt, F. 2020. Modern Cosmology. Academic Press, San Diego.

Einstein, A. 1961. Relativity. The Special and General Theory. Crown Publishing, New York.

Feynman, R. 2011. The Feynman Lectures. Basic Books, New York.

Guex, J., Torday, J.S. and Miller, W.B. Jr. 2020. Morphogenesis, Environmental Stress and Reverse Evolution. Springer, Switzerland.

Guise, T.A. and Mundy, G.R. 1996. Physiological and pathological roles of parathyroid hormone-related peptide. Curr. Opin. Nephrol. Hypertens. 5: 307–315.

Hume, R. and Boyd, G.S. 1978. Cholesterol metabolism and steroid-hormone production. Biochem. Soc. Trans. 6: 893–898.

Kahn, C. 1979. The Art and Thought of Heraclitus: Fragments with Translation and Commentary. Cambridge University Press, London.

Kalra, S. and Priya, G. 2018. Lipocrinology — the relationship between lipids and endocrine function. Drugs Context 7: 212514.

Kasahara, Y., Sato, K., Takayanagi, Y., Mizukami, H., Ozawa, K., Hidema, S., So, K.H., Kawada, T., Inoue, N., Ikeda, I., Roh, S.G., Itoi, K. and Nishimori, K. 2013. Oxytocin receptor in the hypothalamus is sufficient to rescue normal thermoregulatory function in male oxytocin receptor knockout mice. Endocrinology 154: 4305–4315.

Kuhn, T. 1957. The Copernican Revolution. Harvard University Press, Cambridge.

Lawson, E.E., Brown, E.R., Torday, J.S., Madansky, D.L. and Taeusch, H.W. Jr. 1978. The effect of epinephrine on tracheal fluid flow and surfactant efflux in fetal sheep. Am. Rev. Respir. Dis. 118: 1023–1026.

Leng, G., Pineda, R., Sabatier, N. and Ludwig, M. 2015. 60 Years of Neuroendocrinology: The posterior pituitary, from Geoffrey Harris to our present understanding. J. Endocrinol. 226: T173–T185.

Libet, B., Gleason, C.A., Wright, E.W. and Pearl, D.K. 1983. Time of conscious intention to act in relation to onset of cerebral activity (readiness-potential). The unconscious initiation of a freely voluntary act. Brain 106: 623–642.

Lieberman, P. 2009. FOXP2 and human cognition. Cell 137: 800–802.

Mamillapalli, R. and Wysolmerski, J. 2010. The calcium-sensing receptor couples to G alpha(s) and regulates PTHrP and ACTH secretion in pituitary cells. J. Endocrinol. 204: 287–297.

Margulis, L. 1993. Origins of species: acquired genomes and individuality. Biosystems 31: 121–125.

Mazzocchi, G., Aragona, F., Malendowicz, L.K. and Nussdorfer, G.G. 2001. PTH and PTH-related peptide enhance steroid secretion from human adrenocortical cells. Am. J. Physiol. Endocrinol. Metab. 280: E209–E213.

Mehra, J. and Rechenberg, H. 1982. The Historical Development of Quantum Theory, Vol. 1: The Quantum Theory of Planck, Einstein, Bohr and Sommerfeld. Its Foundation and the Rise of Its Difficulties (1900–1925). Springer-Verlag, New York.

Misner, C.W., Thorne, K.S. and Wheeler, J.A. 1973. Gravitation. W.H. Freeman, San Francisco.

Mojzsis, S.J., Arrhenius, G., McKeegan, H.T.M., Nutman, A.P. and Friend, C.R. 1996. Evidence for life on Earth before 3,800 million years ago. Nature 384: 55–59.

Orgeig, S. and Daniels, C.B. 2001. The roles of cholesterol in pulmonary surfactant: insights from comparative and evolutionary studies. Comp. Biochem. Physiol. A Mol. Integr. Physiol. 129: 75–89.

Orgeig, S., Morrison, J.L. and Daniels, C.B. 2015. Evolution, development, and function of the pulmonary surfactant system in normal and perturbed environments. Compr. Physiol. 6: 363–422.

Palomero-Gallagher, N. and Zilles, K. 2019. Differences in cytoarchitecture of Broca's region between human, ape and macaque brains. Cortex 118: 132–153.

Pinheiro, P.L., Cardoso, J.C., Power, D.M. and Canário, A.V. 2012. Functional characterization and evolution of PTH/PTHrP receptors: insights from the chicken. B.M.C. Evol. Biol. 12: 110.

Popper, K. 1959. The Logic of Scientific Discovery. Routledge, London.

Porges, S.W. 1995. Orienting in a defensive world: mammalian modifications of our evolutionary heritage. A Polyvagal Theory. Psychophysiology 32: 301–318.

Purevdorj-Gage, B., Sheehan, K.B. and Hyman, L.E. 2006. Effects of low-shear modeled microgravity on cell function, gene expression, and phenotype in *Saccharomyces cerevisiae*. Appl. Environ. Microbiol. 72: 4569–4575.

Rodman, P.S. and McHenry, H.M. 1980. Bioenergetics and the origin of hominid bipedalism. Am. J. Phys. Anthropol. 52: 103–106.

Romer, A.S. 1949. The Vertebrate Story. University of Chicago Press, Chicago.

Rosen, R. 1958. A relational theory of biological systems. Bull. Math. Biophys. 20: 245–260.

Rowlands, P. 2015. The Foundations of Physical Law. World Scientific Publishing, Singapore.

Rubin, L.P., Kifor, O., Hua, J., Brown, E.M. and Torday, J.S. 1994. Parathyroid hormone (PTH) and PTH-related protein stimulate surfactant phospholipid synthesis in rat fetal lung, apparently by a mesenchymal-epithelial mechanism. Biochim. Biophys. Acta 1223: 91–100.

Rubin, L.P., Kovacs, C.S., De Paepe, M.E., Tsai, S.W., Torday, J.S. and Kronenberg, H.M. 2004. Arrested pulmonary alveolar cytodifferentiation and defective surfactant synthesis in mice missing the gene for parathyroid hormone-related protein. Dev. Dyn. 230: 278–289.

Scerri, E.R. 2019. The Periodic Table: Its Story and Its Significance. Oxford University Press, Oxford.

Schrodinger, E. 1944. What is Life? Cambridge University Press, Cambridge.

Shannon, C.E. 1948. A mathematical theory of communication. Bell System Technical Journal 27: 379–423, 623–656.

Sharara-Chami, R.I., Joachim, M., Pacak, K. and Majzoub, J.A. 2010. Glucocorticoid treatment—effect on adrenal medullary catecholamine production. Shock 33: 213–217.

Spade, P.V. 1999. The Cambridge Companion to Ockham. Cambridge University Press, Cambridge.

Torday, J.S. 2003. Parathyroid hormone-related protein is a gravisensor in lung and bone cell biology. Adv. Space Res. 32: 1569–1576.

Torday, J.S. and Rehan, V.K. 2007. The evolutionary continuum from lung development to homeostasis and repair. Am. J. Physiol. Lung Cell Mol. Physiol. 292: L608–L611.

Torday, J.S. and Rehan, V.K. 2009. The evolution of cell communication: the road not taken. Cell Commun. Insights 2: 17–25.

Torday, J.S. and Rehan, V.K. 2011. A cell-molecular approach predicts vertebrate evolution. Mol. Biol. Evol. 28: 2973–2981.

Torday, J.S. and Rehan, V.K. 2012. Evolutionary Biology, Cell-Cell Communication and Complex Disease. Wiley, Hoboken.

Torday, J.S. 2013. Evolution and Cell Physiology. 1. Cell signaling is all of biology. Am. J. Physiol. Cell Physiol. 305: C682–C689.

Torday, J.S. 2015a. The cell as the mechanistic basis for evolution. Wiley Interdiscip. Rev. Syst. Biol. Med. 7: 275–284.

Torday, J.S. 2015b. A central theory of biology. Med. Hypotheses 85: 49–57.

Torday, J.S. 2015c. Homeostasis as the mechanism of evolution. Biology (Basel) 4: 573–590.

Torday, J.S. 2016. Life is simple-biologic complexity is an epiphenomenon. Biology (Basel) 5: 17.

Torday, J.S. and Miller W.B. Jr. 2016a. Phenotype as agent for epigenetic inheritance. Biology (Basel) 5: 30.

Torday, J.S. and Miller, W.B. 2016b. The unicellular state as a point source in a quantum biological system. Biology (Basel) 5: 25.

Torday, J.S. and Miller, W.B. Jr. 2017. The resolution of ambiguity as the basis for life: A cellular bridge between Western reductionism and Eastern holism. Prog. Biophys. Mol. Biol. 131: 288–297.

Torday, J.S. and Rehan, V.K. 2017. Evolution, the Logic of Biology. Wiley, Hoboken.

Torday, J.S. 2018a. Quantum Mechanics predicts evolutionary biology. Prog. Biophys. Mol. Biol. 135: 11–15.

Torday, J.S. 2018b. From cholesterol to consciousness. Prog. Biophys. Mol. Biol. 132: 52–56.

Torday, J.S. and Miller, W.B. Jr. 2018. Unitary Physiology. Compr. Physiol. 8: 761–771.

Torday, J.S. 2019. The Singularity of nature. Prog. Biophys. Mol. Biol. 142: 23–31.

Torday, J.S. 2020. Consciousness, Redux. Med. Hypotheses 140: 109674.

Torday, J.S. 2021. Life is a mobius strip. Prog. Biophys. Mol. Biol. 167: 41–45.

Torday, J.S. 2022. Hormones and Reality. Elsevier, Amsterdam.

Trivers, R. 2011. The Folly of Fools. Basic Books, New York.

Vermeij, G.J. 2016. Gigantism and its implications for the history of life. PLoS One 11: e0146092.

Whitehead, A.N. 1925. Science and the Modern World. MacMillan, New York.

Whitehead, A.N. 1929. Process and Reality. Macmillan, New York.

Wurtman, R.J. 2002. Stress and the adrenocortical control of epinephrine synthesis. Metab. Clin. Exp. 51: 11–14.

Xu, S., Liu, P., Chen, Y., Chen, Y., Zhang, W., Zhao, H., Cao, Y., Wang, F., Jiang, N., Lin, S., Li, B., Zhang, Z., Wei, Z., Fan, Y., Jin, Y., He, L., Zhou, R., Dekker, J.D., Tucker, H.O., Fisher, S.E., Yao, Z., Liu, Q., Xia, X. and Guo, X. 2018. Foxp2 regulates anatomical features that may be relevant for vocal behaviors and bipedal locomotion. Proc. Natl. Acad. Sci. U S A 115: 8799–8804.

Zalasiewicz,, J., Williams, M., Steffen, W. and Crutzen, P. 2010. The new world of the Anthropocene. Environ. Sci. Technol. 44: 2228–2231.

Zhang, F., Wen, Y. and Guo, X. 2014. CRISPR/Case9 for genome editing: progress, implications and challenges. Hum. Mol. Genet. 23: R40–R46.

Zheng, W., Wang, Z., Collins, J.E., Andrews, R.M., Stemple, D. and Gong, Z. 2011. Comparative transcriptome analyses indicate molecular homology of zebrafish swimbladder and mammalian lung. PLoS One 6: e24019.

Chapter 18
The Cell as a Mobius Strip

Introduction

Louis Kauffman the mathematician tells us that if bisect a mobius strip, its edges form a Trefoil Knot, which must be reducible to a circle if it is a true mathematical knot. The cell is a homolog of the mathematical knot since it must also be able to unknot itself (meiotically) to form the egg or sperm in order to reproduce.

The interrelationship between the knot and the cell is thought-provoking in biological terms given that the Trefoil Knot could be considered a homolog for the three primary germ layers, namely the endoderm, ectoderm and mesoderm, that interact to generate the multicellular embryo, beginning with the zygote.

But why would there be an interrelationship between a mathematical knot and a cell? This seems to be totally counterintuitive, and yet they are both circular in two dimensions, and spherical in three dimensions. In the case of the cell, the sphere is the most efficient physiologic form for metabolism because metabolically it is the optimal ratio of surface area for gas exchange. In the case of the geometric sphere, Schwarzschild's Radius is the mathematical reduction of Einstein's Field Theory to a Black Hole. The lipid barrier generated by micelles forms a 'hole' in the infinite plane of the Implicate Order; by homology, a Black Hole forms a hole in the Cosmos. Cosmic fibers course through the Cosmos much the same way that the cytoskeleton provides form to the cell. Functionally, the cytoskeleton determines whether the cell is homeostatic, meiotic or mitotic, controlled by Target of Rapamycin signaling.

The cell membrane as the basis for the mobius strip

Upon further reflection, the cell membrane is like a mobius strip because it forms a continuous topology between the inner and outer environments of the cell, initially giving rise to the 'inside' and 'outside' when micelles first formed. Stated otherwise, when lipids were first immersed in water they spawned the 'concept' of the mobius strip, because prior to that inside and outside did not exist—there was only one infinite plane in the Implicate Order. Interestingly, the mobius strip resembles the symbol for infinity.

To gain an understanding for this homology, go back to the physicochemical origin of the cell as amphiphilic lipid molecules floating on the surface of the primordial waters that covered the Earth some 100 million years after its formation. The negatively-charged hydrophilic ends of the lipid molecules pointed downward into the water, the positively-charged ends pointed upward due to the gravitational force of the Earth. When adequate amounts of lipid molecules aligned packed together to reduce the surface tension of the water surrounding themselves, they neutralized the Van Der Waals Forces that generate the surface tension of water, spontaneously forming micelles, or lipid spheres. Those micelles formed the Explicate Order as distinguished from the Implicate Order, as described in David Bohm's "Wholeness and the Implicate Order". In other words, the pre-adaptation for micelles was the lipid molecules immersed in water, pre-existing "inside and outside"; inside and outside only came about once the micelles emerged from those lipid molecules. And please bear in mind that the transition from individual lipid molecules to micelles constituted a quantum change. Therefore, the micelle was the origin of the mobius strip.

The hypothesis being tested is that the lipid barrier of the Explicate Order 'conceptualized the mobius strip' by dividing the Implicate Order into 'inside and outside', conceiving life in the process. That is the origin of consciousness as the First Principles of Physiology, referencing the Singularity that pre-dated the Big Bang.

Micelles, semi-permeable membranes and calcium fluxes

The semi-permeable membranes of micelles allow particles to enter and exit as a function of osmotic pressure, forming the basis for the protocell. Calcium ions were prevalent in the primordial ocean, dissolved from the bedrock. As the carbon dioxide produced by plants accumulated in the atmosphere, it dissolved in the waters below, producing carbonic acid, hastening the dissolution of the bedrock, releasing ever-more calcium into the waters.

The accumulation of calcium ions within micelles was sped-up by the Sun warming them by day, causing the lipid membranes to liquify, thereby expanding; at night, the micelles would cool and re-conform to their original size and shape due to hysteresis, or 'molecular memory'. That recursive expanding and contracting caused a build-up of calcium ions within the protocells. But since calcium ions are toxic to lipids, causing them to denature, as a consequence a sub-set of micelles evolved the capacity to control the entry and exit of calcium ions mediated by calcium channels.

The flow of calcium ions through the cell constitutes the biologic flow of energy. Experimentally, if cells are exposed to zero gravity they lose their capacity to generate a calcium ion flow. This critical role of gravity in the formation and initiation of life even applies to abiotic micelles. The causal relationship between the force of gravity and the 'life force' epitomizes the intimate relationship between physics and biology. That interrelationship transcends the cell because Einstein's Field Theory states that when gravity impinges on a curved surface it produces energy. The principle of Quantum Entanglement would have entrained particles within the micelle, stabilized by such gravity-dependent energy. Based on Quantum Mechanical principles, the configuration of such entangled particles would have referenced non-local events in the Cosmos. This foundation for the micelle would have provided the pre-adaptive condition for the emergence of symbiogenesis—the acquisition of factors in the environment that have posed existential threats over the course of the 'history' of the organism, or what we commonly refer to as cellular evolution. In fact, our physiognomy is that history, and it is why we carry it along as our long-term memory, warts and all.

The evolution of multicellular organisms

Over the course of cellular evolution, there was a point at which prokaryotes (bacteria) devised ways of imitating multicellularity, namely Quorum Sensing and Biofilm, providing a biologic advantage over their unicellular eukaryotic competitors. In order to survive, eukaryotes devised means of expressing authentic multicellularity using cell-cell communications based on soluble growth factors signaling to their cognate receptors on neighboring cells.

Beginning with the zygote, the animal and vegetal poles intercommunicate using such growth factors to initiate and perpetuate cell division and differentiation that ultimately terminate as the offspring. Beyond birth, the same cell-cell signaling mechanisms maintain homeostasis within and between cells, constituting the organism. A break-down in cell-cell communications leads to remodeling of structure and function, constrained by homeostasis. Physiologic stresses can cause disruption of homeostatic control, leading to the production of Radical Oxygen Species (ROSs); ROSs can cause gene mutations and duplications, offering the opportunity for variations in cell-cell signaling to accommodate such genetic changes. Once a new homeostatic set-point is reached, the cells involved will maintain and perpetuate that state adaptively in what we conventionally recognize as evolution.

For example, the vertebrate lung evolved from the swim bladder of boney fish, both of which are 'gas-exchangers'. In the case of the former, the swim bladder utilizes gases to facilitate efficient feeding at different depths of water, using such gases to inflate or deflate the bladder for buoyancy, optimizing metabolism by minimizing energy output. In the case of the latter, the lung facilitates the uptake of oxygen for metabolic activity within the tissues and organs of the body, more directly achieving the same ends as the swim bladder—same genes, different purposes, or what is descriptively referred to in biology as pleiotropy.

This narrative for the emergence of vertebrates from water to land exemplifies the 'long-game' of evolution. Carbon Dioxide in the atmosphere accumulated due to the evolution of plant life on Earth. The CO_2 in the atmosphere caused a 'greenhouse effect' (The Romer Hypothesis), warming the ambient temperature of the atmosphere. That process had a two-fold effect on life, reducing the amount of oxygen dissolved in the water due to the partial pressure of gas being dependent on the water temperature, and in tandem partially drying up the water covering the surface of the Earth, exposing land masses for potential subsequent habitation. The combined effect resulted in a subset of boney fish moving out of water onto land. It was the so-called physostomous species of boney fish that formed the basis for land vertebrates, having a pneumatic duct that connects the esophagus to the swim bladder, forming that structural homology between fish and land vertebrates.

Tiktaalik is the fossilized evidence for this water-land transition, representing the transition from fins to legs. The 'fossil record' for the evolution of the internal organs is more difficult to ascertain since there is no 'hard' fossil evidence. However, there are cellular-molecular data embedded in ontogeny and phylogeny that document the history of the organism, and cell-cell signaling allows recalling such events.

The fossil skeletal evidence indicates that vertebrates attempted to breech land on at least five separate occasions, inferring a step-wise "ratcheting-up" process for evolving from the swim bladder to the lung, for example. The final common biomolecular pathway for this series of events is lung surfactant, the 'soapy' material that prevents the alveoli from collapsing (atelectasis) under the hydrostatic force of the water tension caused by the fluid lining the alveoli. Surfactant is also present in the swim bladder, where it prevents the walls of the bladder from adhering. The bladder evolved into alveoli through a progressive increase in the surface area of the gas-exchange chamber, effectively enhancing the surface area-to-blood volume ratio for oxygenation. But for that to occur, there had to have been tandem increases in the amount and/or quality of lung surfactant since surface tension is inversely related to the diameter of a sphere based on the Law of Laplace. As the lung evolved, both of these properties of lung surfactant have been documented to have changed commensurately to one degree or another, averring the causal interrelationships.

Over the course of this challenge to adapt to land other physiologic systems evolved in tandem, including the endocrine system. As fish evolved into quadrupeds, emerging from adaptive buoyancy in water facilitated by the swim bladder to the lung for air breathing in adaptation to land life, there was a concomitant increase in the effective force of gravity on the organism. As a result, there were three documented receptor gene duplications—the Parathyroid Hormone-related Protein Receptor (PTHrPR), the Glucocorticoid Receptor (GR), and the ßAdrenergic Receptor (ßAR). All three of these receptor gene duplications were necessary for the evolution of the lung, since they all resulted in amplification of the signaling pathways for physiology that they determine. The PTHrPR was existential for the formation of alveoli since its deletion results in death at birth in mice; the ßAR was essential for the evolution of the pulmonary circulatory system independently from the systemic circulatory system; the GR was vital for the increase in the ßAR receptor density within the aforementioned lung circulatory system. The selection pressure for the PTHrPR may have initially been due to its role in the calcification of bone, strengthening the skeletal system in order to support the organism's increased weight on land; and again, selection pressure for the GR evolved from the increase in blood pressure on land due to gravity, causing shear stress on the microcirculation, which would have been offset by the evolution of the GR from the mineralocorticoid receptor, epistatically off-setting the effect of the latter on blood pressure. The advent of the GR would have facilitated the distribution and number of ßARs, particularly allowing for the independent regulation of the lung circulatory system, as mentioned above.

On the role of epigenetic inheritance for evolution

Seen from this foundational cellular-molecular perspective, it is of value to consider the mechanism of epigenetic inheritance. Elsewhere, it has been hypothesized that the phenotype is not merely the physical form of the organism, but is actually the means by which it expresses 'agency' in order to effectively monitor the environment for changes that might prove existentially life-threatening. By assimilating such changes in the egg and sperm as biochemical adducts of DNA that modify its readout, the offspring are enabled to adapt to such environmental changes.

Conclusions

In this way, the cell is able to perpetuate itself, defined by the cell membrane structurally and functionally distinguishing the outside of the cell from the inside. As the origin of the mobius strip, the cell membrane 'remembers' its history, referring all the way back to those lipids originating in deep space, orienting themselves vertically to the surface of the waters, pointing upward to the Sun, conducting thermal energy to fuel life. The innate, holistic connection between the lipid origin of life and the topology of the mobius strip attest to the Singularity of Nature. Elsewhere it has been postulated that the unicell is actually the primary state of being, and that its purpose is to remain in close proximity to the Singularity that existed prior to the Big Bang. The idea that the Explicate Order formed from the infinite of the Implicate Order as the basis for the mobius strip as both inside and outside is consistent with that perspective.

Viewing this phenomenon retrospectively is misleading. For example, Hoffmeyer has also deduced that the cell membrane behaves like a mobius strip, yet he makes the same error in judgement that Knot Theoreticians and Rewrite Mathematicians do by thinking of such processes only in the 'now', synchronically, whereas the basis for these interrelationships must be seen diachronically 'across space-time' from their origin ontologically in order to recognize the primacy of the flow of energy over the material epiphenomena. By way of explanation, it is like Husserl saying that the origin of geometry is not innate to our being, the geometric figures being projections of our acquired knowledge of such forms from the Cosmos, like Plato's Theory of Forms being archetypes in our minds, or 'shadows' on cave walls. So, too, are mathematical 'knots' archetypes, which must be unknottable as circles. Such circles are homologs of the unicell, which also knots after a fashion developmentally to form complicated organisms, but must also unknot during meiosis to form the egg and sperm. It is noteworthy that it has been discovered that there are 'fibers' in the Cosmos that course through the heavens, much like the cytoskeleton, which determines the physiologic state of the cell. The cytoskeleton could hypothetically be queried as a 'bioassay' for the Cosmological Fibers in much the same way that Smolin has expressed the organic nature of the Cosmos.

And Peter Rowlands' Rewrite Math is homologous with biologic evolution, the cell being the equivalent of 'zero' as an attractor, and the introduction of a new value to the Rewrite Math requiring an evaluation of the existing set of values. The same occurs in epigenetic evolution, new epigenetic data having to be evaluated by the genome of the egg and sperm before it can be assimilated.

In short, the above comparison and contrast between induction and deduction is differentiated by the judicious use of experimental data to leverage knowledge, moving ever closer to Bohm's Implicate Order, but never quite reaching it since we have evolved out of ambiguity and must remain in that state as our *raison d'etre*. In so doing, we are not only able to understand Heisenberg 'uncertainty', we live by its principle as the embodiment of the atom. This state of being also would explain Gödel's 'Incompleteness Theorem'.

Chapter 19
Two Paranormal Psychologists Walk into a Bar

Mossbridge and Baruss

In her discussion about her work in paranormal psychology, Julia Mossbridge tells Robert Lawrence Kuhn, the host of "Closer to Truth" that the world begins with consciousness (https://www.youtube.com/watch?v=rbhPP4Pdx24), and that consciousness has formed everything we think of as reality. Don Hoffman is similarly inclined, referring to this as a 'miracle', or a given. In a lengthy discussion with Lex Fridman, he proposes that neurons were built from consciousness, to which Fridman reacts strongly, saying that 'blows his mind'. Yet I am of the opinion that neurons evolved from consciousness as a consequence of Symbiogenesis (Sagan, 1967), the mechanism by which factors in the environment outside of the cell that have posed existential threats have been assimilated, compartmentalized and linked together functionally to form our physiology (Torday and Rehan, 2012; Torday and Rehan, 2017).

Of course, that presupposes the existence of the cell, forming a cell membrane barrier between the environment and the interior of the cell (Torday, 2022). Chapter 12 described the process by which lipids originating in deep space formed the cell membrane due to the force of the Earth's gravity orienting the negatively-charged ends of these amphiphiles downward into the water, being negatively charged; when a critical mass of such lipid molecules packed together the cumulative negative charge 'neutralized' the Van der Waals Force for surface tension of water. That allowed the lipid molecules to 'leap' into micelles 'quantumly'. Subsequently, the components of the cell's physiology were acquired through the mechanism of Symbiogenesis. That process formed a 'simulacrum' of the Laws of Nature within the cell, providing the common denominator with the Cosmos, whose 'logic' (Torday and Rehan, 2017) derives from stellar nucleosynthesis, the generation of starlight producing the Elements in the sequence of their atomic mass. The assimilation of the Elements infers that same logic within the cell. Thus, the basis for the consciousness that Mossbridge and Baruss reference in their work is derived from that process.

The role of gravity in the process of life began with the Big Bang forming the four fundamental forces—the strong force, the weak force, electromagnetism and gravity. It is thought that the force of gravity formed the planets (Zoe et al., 2005), and subsequently formed micelles from the lipids carried by the snowball-like asteroids that pummeled the atmosphere-less Earth. There too, Earth's gravity was the 'lodestone' for those asteroids. If we then fast-forward to experimental evidence showing that gravity is necessary for consciousness (Torday, 2023), there is a '(w)holism' surrounding the relationship between life on Earth and gravity. In turn, that observation of the fundamental role played by gravity for consciousness, both the local and the non-local, provides insight to our origin in the consciousness of the Cosmos (Torday, 2023) referred to by Mossbridge and Baruss. They acknowledge that based on intuition or deduction, whereas the current depiction for consciousness is solely founded on experimental evidence, providing the means for connecting to other physiologic aspects of consciousness in a testable and refutable context.

And just to be clear, the above could be construed as panpsychism—that everything is conscious—but that would be in error. Consciousness as we currently understand it is an anthropomorphism for that feeling of something greater than ourselves, being the net result of Symbiogenesis.

Furthermore, the concept of 'unconsciousness' becomes more relevant to understanding consciousness. Given that we evolved from nocturnal rodent-like organisms into diurnal organisms, some of the 'old ways' remain within our physiology. So the distinctions between waking and dreaming are arbitrary with regard to consciousness. Indeed, during Rapid Eye Movement sleep when we have dreams norepinephrine is involved in their processing (Becchetti and Amadeo, 2016). And similar processing mediates the 'fight or flight' mechanism (Dienstbier, 1989), versus problem solving. The inference here is that there is a common motif for sorting out 'reality' that is mediated by the same biochemical mechanism. So for example, when we resolve problems by 'sleeping on them' we are utilizing a sleep-related mechanism that may find its way into our consciousness.

Perhaps this interrelationship forms the basis for 'lucid dreaming'. Or the Runner's High, due to endorphins (though that has been disputed), putting the individual into a dream-like state while highly awake. Or Near Death Experiences, in which stress brings on a flood of endocrine hormones, potentially causing a 'wave collapse' that puts us in touch with our unicellular origin? Given that our physiology is derivative of the latter? It is in the unicellular state, with only our semi-permeable cell membrane between us and the Cosmos that we are the most proximate to the Implicate Order.

Paul Klee's Esthetic theory of life and gravity

All of the above is encountered to one extent or another in Paul Klee's "Notebooks, Volume 2, The Nature of Nature" (1970), in which he relates his experience as an artist expressing 'the world'. Similarly, in Michael Spitzer's "A History of Emotion in Western Music" (2020), not unlike the interrelationship of evolution and history (Torday, 2019), the author traces music as feelings on the part of the composer and the listener alike. Likewise, in "Hormones and Reality" (2022) I show how and why our endocrine system, under the auspices of epigenetics, has guided our evolutionary history (Zhang and Ho, 2011).

Beyond all of that, Klee's conceptualization of how he transfers life to canvas is reminiscent of Spencer-Brown's "Laws of Form", making a 'mark' on an otherwise unmarked surface, recapitulating the origin of life as a break in the surface tension of water. And then there's Stafford Beer's theory of 'the firm', having to be connected to 'the pond' as Nature, acknowledging the fundamental interrelationship between it and man's 'enterprise' (Beer, 1972).

Einstein's theory of gravity

In contrast to Newton's view of gravity as the attraction of bodies, Einstein's view of gravity in his 'Special Relativity' is as the distortion of the fabric of space-time. It was that Einsteinian view of gravity that was 'neutralized' in the experiment where the phenotype of the cell was lost (Torday, 2023), perhaps entering Klee's primordium (1970):

"Chaos and anarchy, a turbid jumble. The intangible—nothing is heavy, nothing light (light-heavy); nothing is white, nothing black, nothing red, nothing yellow, nothing blue, only an approximate grey."

So there is a paradox: in the article "Life is a mobius strip" (Torday, 2021), I made the case for the primordial micelle having a mobius strip-like cell membrane that has neither an inside nor an outside..... functionally. That affords the cell the 'imagination' to remember back to when it wasn't. If that's the power that allows the cell to imagine time travel, is that real or virtual? The author is inclined to think it is virtual since real time travel is currently not scientifically testable.

References cited

Becchetti, A. and Amadeo, A. 2016. Why we forget our dreams: Acetylcholine and norepinephrine in wakefulness and REM sleep. Behav. Brain Sci. 39: e202.

Beer, S. 1971. Brain of the Firm, Beer Allen Lane, London.

Dienstbier, R.A. 1989. Arousal and physiological toughness: implications for mental and physical health. Psychol. Rev. 96: 84–100.

Klee, P. 1970. Notebooks, Volume 2, The Nature of Nature. Schwabe&Co., Basel.

Leinhardt, Z.M. and Richardson, D.C. 2005. Planetesimals to Protoplanets. I. Effect of fragmentation on terrestrial planet formation. Astrophysical J. 625: 427–440.

Sagan, L. 1967. On the origin of mitosing cells. J. Theor. Biol. 14: 255–274.

Spitzer, M. 2020, A History of Emotion in Western Music. Oxford University Press, Oxford.

Torday, J.S. and Rehan, V.K. 2012. Evolutionary Biology, Cell-Cell Communication and Complex Disease. Wiley, Hoboken.

Torday, J.S. and Rehan, V.K. 2017. Evolution, the Logic of Biology. Wiley, Hoboken.

Torday, J. 2019. Evolution, the mechanism of big history—The Grande synthesis. J. Big History III: 17–24.

Torday, J.S. 2021. Life is a mobius strip. Prog. Biophys. Mol. Biol. 167: 41–45.

Torday, J.S. 2022. Evolution, gravity, and the topology of consciousness. Prog. Biophys. Mol. Biol. 174: 50–54.

Torday, J.S. 2023. Consciousness, embodied Quantum Entanglement. Prog. Biophys. Mol. Biol. 177: 125–128.

Zhang, X. and Ho, S.M. 2011. Epigenetics meets endocrinology. J. Mol. Endocrinol. 46: R11–R32.

Index

A

agent 9, 12, 18, 54, 61, 70, 81, 83, 88, 95, 98, 100, 106, 109, 113, 122, 129, 131, 136, 151, 156
algorithm 80, 86, 88, 100
ambiguity 28, 42, 74, 82, 99, 114, 132, 141, 167
amphiphile 16, 36, 59, 62, 66, 68, 81, 98, 168
Andy Clark 114, 132, 157
Area of Broca 1, 34, 37, 140, 147
art 2, 37, 38, 56, 60–62, 71, 101
asteroid 16, 42, 45, 59, 68, 76, 81, 98, 118, 169
autopoiesis 47
awareness 6, 60, 63, 94, 127
Baruss 168, 169
bauplan 33

B

βAdrenergic Receptor 34, 35, 48, 137, 165
Bekenstein 56, 58
Big Bang 7, 8, 10–13, 17, 26, 40–43, 70, 78, 82, 87, 99, 104–106, 108–110, 116, 141, 143, 144, 148, 149, 153–157, 162, 166, 169
Biofilm 8, 36, 105, 163
bipedalism 1, 29, 34, 35, 37, 40, 50, 54, 89, 137, 140, 146, 147, 151
bird 6, 10, 18, 67, 69, 70, 81, 98, 107, 115, 121, 133, 138, 139, 147
Black hole 8, 56, 58, 104, 161
Bohm 9, 10, 33, 39, 82, 90, 95, 96, 99, 106, 124, 141, 142, 152, 154, 162, 167
bone 2, 20, 21, 27, 37, 48, 49, 63, 120, 129, 130, 137–140, 154, 155, 165
boney fish 2, 48, 75, 86, 137, 164
brain 1, 3, 4, 11, 25, 26, 29, 34, 40, 49, 51, 54, 67, 74, 83, 89, 100, 107, 108, 112–115, 119, 123, 124, 127, 130–133, 148, 151

C

calcium flux 35, 48, 49, 130, 142, 154, 155, 162
Candace Pert 113, 131
carbon dioxide 30, 36, 43, 90, 137, 162, 164
cell 2, 3, 5–9, 11–13, 16–19, 21–23, 25–27, 29, 30, 33–45, 47–53, 56–63, 65–71, 75–83, 86–91, 94–100, 103–106, 108, 110, 112–114, 116–126, 128–132, 136, 138, 140–144, 147–149, 151, 153–156, 161–164, 166, 168–170
cell-cell communication 2, 3, 7–9, 16, 17, 22, 27, 30, 33–36, 41, 44, 51, 52, 62, 63, 66, 70, 71, 75, 78, 83, 89, 94, 96, 100, 103, 105, 106, 113, 119, 124, 131, 142, 143, 154, 156, 163
central nervous system 1, 11, 29, 37, 40, 50, 54, 89, 107, 115, 133, 147
Chalmers 3, 5, 35, 54, 83, 94, 114, 132, 157
chemiosmosis 6, 17, 36, 38, 43, 83, 95, 100, 117, 118, 128
cholesterol 16, 27, 30, 35, 36, 39, 41, 69, 76, 81, 87, 98, 119–123, 125, 147
cipher 17, 90
Claude Bernard 59, 66, 96
coherence 10, 42, 83, 100, 106, 107, 129
consciousness 1, 3–6, 9, 29, 30, 35, 41, 44, 51–54, 57, 59–63, 74, 76–79, 83, 86, 88, 90, 91, 94, 106, 112–115, 125–133, 136, 140, 142, 153–155, 157, 162, 168, 169
cosmos 7–11, 17, 22, 23, 27–30, 43–45, 50, 52, 56, 58–63, 66, 68, 71, 77, 78, 81, 83, 86, 88, 90, 94, 95, 100, 103, 104, 106, 108, 112, 114, 125, 128–130, 132, 136, 141–143, 149, 153–156, 161, 163, 166, 169, 170
cytoskeleton 70, 82, 87, 88, 99, 154, 161, 166

D

Darwin 10, 34, 52, 63, 91, 108, 141
deconvolute 52, 65, 94
Descent of Man 33, 44
developmental biology 138, 139, 143
diachronic 6, 21, 22, 31, 36, 88, 116, 124, 142–144, 148, 155
DNA 8, 17, 22, 27, 49, 55, 61, 67, 68, 77, 79, 80, 83, 88, 90, 97, 100, 105, 113, 131, 151, 165
Don Hoffman 168
Double-Slit Experiment 113, 130

E

Earth 2, 6–10, 13, 16, 18, 41, 43, 45, 53, 58, 59, 66, 68, 75, 76, 78, 81, 86, 90, 95, 96, 98, 104–107, 109, 117, 118, 129, 130, 148, 152, 153, 162, 164, 169
egg 9, 12, 18, 22, 28, 47, 49, 50, 57, 61, 77, 80, 83, 95, 100, 106, 108, 113, 114, 119, 131, 132, 143, 149, 161, 165, 166
Einstein 17, 27, 30, 53, 63, 87, 103, 129, 141, 144, 152, 161, 163, 170
element 9, 21–23, 42, 43, 45, 77, 106, 116, 117, 139, 141–145, 148, 150, 151, 153, 155, 169
embryo 9, 28, 88, 106, 119, 120, 123, 161
emergence 12, 16, 17, 27, 28, 30, 40, 41, 48, 52, 53, 74, 78, 108, 122, 125, 163, 164
endocrine system 1, 41, 60, 83, 113, 131, 142, 147, 165, 170
endosymbiosis 8, 41, 43, 45, 96
energy 2, 6–13, 17, 18, 20, 21, 26, 27, 29–31, 33, 37, 38, 40, 42, 43, 53, 56, 58, 59, 62, 63, 67, 69, 76, 78, 79, 81, 83, 87–89, 94–96, 98, 103–110, 116–118, 128, 129, 141, 147, 149, 151–157, 163, 164, 166
environment 2, 5, 6, 8, 9, 12, 16, 18, 21, 22, 24, 25, 27–30, 36, 41–45, 50, 53–55, 57–59, 62, 63, 67–69, 77, 79, 82, 87, 88, 90, 95–97, 99, 105, 106, 109, 113, 114, 116, 117, 119, 123, 125, 128, 131, 132, 136, 143, 146, 147, 149, 150, 152–154, 156, 162, 163, 165, 168
epigenetic 1, 6, 7, 9, 12, 16, 18, 22, 23, 34, 36, 43, 47, 49, 50, 52, 54, 55, 57, 58, 61, 67, 77, 80, 83, 88, 90, 91, 95, 97, 100, 103, 106, 108, 109, 113, 114, 122–124, 131, 132, 144, 149, 152, 165, 166, 170
epigenetic mark 6, 12, 18, 43, 50, 88, 108, 113, 122, 123, 131

epistemology 62, 70, 100, 114, 127, 132
eukaryote 40–42, 66, 96, 118, 121, 122, 130, 155, 156, 163
evolution 1–4, 6–13, 16–18, 20–23, 25–31, 33–45, 47–54, 56–59, 61–63, 65–71, 75–82, 84, 86, 87, 89–91, 94–101, 103–110, 112–125, 130, 131, 137, 138, 140, 142–152, 155–157, 163–166, 170
evolutionary biology 141–143, 148, 150
exaptation 10, 16, 41, 67, 68, 70, 79, 81, 82, 97–99, 107, 118
Explicate order 33, 36, 38, 39, 59, 61, 90, 95, 129, 141, 142, 151, 152, 154, 162, 166

F

falsifiable 36, 91, 141
Fibonacci 47, 49, 50, 56, 58, 60, 80
Field theory 27, 53, 63, 129, 141, 161, 163
fight or flight 54, 60, 146, 169
First Principles of Physiology 2, 6, 7, 17, 34, 36, 38, 42, 43, 54, 76, 83, 86, 95, 100, 103, 113, 116, 117, 128, 147, 152, 162
fish 2, 10, 27, 40, 48, 65, 67, 69, 70, 75, 81, 86, 94, 98, 107, 121, 137, 139, 140, 143, 144, 148, 164, 165

G

gastrulation 28, 47, 151
germ cell 22, 55, 57, 58, 61, 80, 95, 123
Glucocorticoid Receptor 48, 91, 165
Godel 167
gravitropism 74
gravity 2, 10, 16, 17, 20, 21, 26, 27, 29, 30, 34, 36, 48–50, 53, 56, 58, 59, 62, 63, 67, 74–76, 78, 81, 90, 91, 94, 95, 98, 107, 112, 126–130, 141, 151, 154, 155, 163, 165, 168–170
Greenhouse effect 2, 20, 30, 48, 54, 90, 137, 164
Guex 67

H

hard problem 3, 5, 35, 94, 114, 132
Heart 26, 29, 69, 70, 74, 81, 82, 98, 151
History 2, 3, 12, 13, 16, 22, 24, 28–30, 33, 43, 44, 55, 67, 86–88, 94, 109, 113, 114, 117, 118, 122, 124, 126, 127, 131, 163, 164, 166, 170

holism 2, 3, 6, 33, 40, 50, 71, 86, 90, 100, 110, 155, 169
homeobox gene 26, 68, 80, 97
homology 10, 11, 37, 43, 57, 66, 67, 70, 82, 86, 95–97, 99, 107, 139, 143, 144, 148, 156, 161, 162, 164
hydrophilic pole 17, 53
hydrophobic pole 53

I

Implicate order 10, 33, 36, 39, 59, 61, 66, 70, 81, 82, 90, 96, 98, 99, 106, 129, 141, 142, 151, 152, 156, 161, 162, 166, 167, 170
inflection point 33

J

Just So Stories 114, 132, 141

K

kidney 2, 27, 29, 49, 69, 70, 81, 98, 108, 124, 137, 143, 148, 154
Knot Math 50, 57

L

lamprey 24, 25, 77
language 1, 19, 29, 33–41, 44, 45, 50, 54, 89, 140, 146, 147
Law of Laplace 165
Laws of Nature 9, 27, 88, 106, 125, 143, 152, 153, 169
left-right brain 29, 74, 83, 100
Libet Experiment 4, 51, 112, 130, 142
life cycle 18, 44, 57, 66, 67, 80, 83, 91, 96, 97, 100, 119, 122, 149
lipid molecules 2, 19, 30, 36, 40, 49, 53, 56, 59, 63, 76, 78, 81, 94, 95, 104, 127, 128, 162, 168
local 3, 5, 17, 22, 27, 29, 33, 43, 50, 51, 58–63, 78, 87, 103, 112, 114, 125, 126, 128–130, 132, 140, 141, 163, 169
Logic 22, 23, 28, 45, 74, 77, 118, 136, 142, 143, 153, 156, 169
Louis Kauffman 161
lucid dreaming 170
lung 2, 7, 10, 11, 16, 18, 20, 21, 25, 27, 33, 34, 37, 40, 43, 48–50, 52, 54, 63, 69, 70, 75, 81, 82, 86, 89, 91, 98, 99, 104, 107, 108, 120, 121, 124, 125, 129, 130, 132, 137, 143–148, 154–156, 164, 165

M

Maslow Peak Experience 44, 60, 62, 112, 126, 130, 142
mathematics 27, 30, 50, 56–59, 61, 80
Maturana 47
Meckel's cartilage 137–139
Meditation 3, 51, 60, 62, 89, 112, 130, 140
memory 2, 3, 16, 39, 42, 49, 51, 54, 55, 67, 76–79, 90, 97, 113, 114, 131, 132, 163
mesenchymal-epithelial interaction 65
metabolism 27, 30, 35, 37–41, 50, 57, 75, 87–89, 122, 123, 125, 146, 161, 164
Mobius strip 18, 29, 30, 50, 53, 58, 60, 61, 77, 78, 142, 161, 162, 166, 170
Mossbridge 168, 169
music 37, 56, 60, 82, 99, 136, 170

N

N-Acetylcholine Receptor 20, 21, 31
Nature 3, 9, 10, 12, 17, 20, 21, 27, 33, 39, 44, 48, 50, 51, 61, 63, 70, 71, 74, 78, 82, 88, 95, 99, 100, 106–108, 113, 114, 121, 124–126, 129, 130, 132, 137, 142, 143, 145, 149, 150, 152, 153, 157, 166, 169, 170
Near Death Experience 3, 44, 60, 62, 112, 126, 130, 142, 170
negative entropy 2, 6, 17, 38, 74, 82, 83, 88, 95, 99, 100, 128, 141, 147, 149, 154
Newton 104
Noise 136
non-local 3, 5, 17, 22, 27, 29, 33, 43, 50, 51, 58–63, 78, 87, 103, 112, 114, 125, 126, 128–130, 132, 140, 141, 163, 169

O

Occam 6, 7, 10, 18, 103, 156
Offspring 9, 12, 17, 18, 22, 29, 49, 57, 60, 61, 78, 79, 88, 95, 106, 108, 109, 113, 123, 131, 143, 163, 165
ontogeny 7, 8, 17, 20, 21, 24, 25, 34, 67, 75, 80, 91, 94, 97, 103–105, 114–117, 124, 128, 132, 137, 164
Out of Body Experience 44, 60, 62, 78, 112, 126, 130, 142
oxygen 2, 6, 16, 20, 21, 25, 30, 34, 36, 39, 40, 43, 48, 57, 68, 69, 76, 80, 81, 87, 90, 97, 98, 116, 120, 125, 145, 147, 156, 163, 164

P

paranormal 168
Parathyroid Hormone-related Protein 11, 18, 25, 34, 40, 48, 63, 67, 108, 125, 130, 137, 144, 145, 165
Pauli Exclusion Principle 42, 43, 83, 100, 143
Periodic Table of Elements 22, 45, 141–143, 148, 150, 155
phanerozoic 76
phenotype 5, 8–11, 13, 16, 17, 23, 26, 49, 54, 76, 91, 105, 106, 108, 110, 117, 120, 122, 124, 126, 129, 143, 150, 151, 156, 165, 170
phylogeny 6–8, 13, 17, 20, 21, 34, 53, 67, 70, 74–76, 80, 94, 95, 97, 100, 103–105, 109, 114–117, 121, 124, 128, 132, 137, 143, 150, 152, 155, 164
physics 13, 18, 20, 27–29, 41–43, 56, 58, 60–62, 83, 100, 110, 124, 127, 129, 143, 144, 149, 154, 163
physoclistous 75
physostomous 48, 164
planet 10, 12, 26, 45, 59, 90, 107, 109, 169
pleiotropic 21, 31, 43
probabilistic 12, 42, 43, 83, 100, 108, 127, 143
protocell 26, 28, 39, 49, 50, 53, 58, 59, 63, 77, 87, 94, 95, 118, 141, 162, 163

Q

quadruped 165
Quantum entanglement 3, 5, 17, 22, 26–28, 33, 50, 53, 59, 60, 62, 63, 67, 76–78, 87, 95, 103, 112, 124, 125, 127–129, 141, 155, 157, 163
Quantum mechanics 3, 11, 13, 16, 17, 27–30, 42, 43, 52, 53, 61, 63, 67, 83, 84, 87, 88, 91, 95, 100, 108, 110, 125, 127, 141
Quorum sensing 8, 36, 105, 163

R

Radical Oxygen Species 57, 68, 69, 97, 98, 163
random mutation 13, 16, 65, 86, 88, 94, 110, 116
Rapid Eye Movement 67, 169
redshift 7, 74, 104

reptile 69, 70, 81, 98, 121, 138, 139
reverse engineer 124
Rewrite Math 56, 57, 80, 166
RNA 8, 105, 154
Romer 2, 48, 54, 90, 164
Rowlands 56, 57, 80, 145, 149, 152, 157, 166
Runner's High 3, 60, 62, 78, 112, 126, 130, 142, 170

S

Sagan 22, 45, 53
Schrodinger Wave 136
Schwarzschild's Radius 161
science 13, 26, 37, 43, 61, 110, 114, 132, 141, 149
Scientific Method 13, 61, 110, 127, 136
self-organization 42, 47, 49, 50, 119
self-reference 47, 116
semi-permeable membrane 17, 27, 59, 66, 69, 90, 98, 129, 162
simulacrum 56, 57, 169
singularity 7, 8, 11, 20, 41, 43, 44, 70, 71, 82, 87, 99, 100, 104, 105, 108, 148, 149, 152–155, 157, 162, 166
Smolin 8, 41, 104, 166
soluble growth factor 3, 7–9, 13, 16–18, 22, 35, 36, 45, 47, 62, 63, 79, 89, 103, 105, 109, 119, 142, 156, 163
sperm 9, 12, 18, 22, 28, 49, 50, 57, 61, 77, 80, 83, 95, 100, 106, 108, 113, 114, 131, 132, 161, 165, 166
Stafford Beer 170
Stellar nucleosynthesis 9, 21–23, 77, 106, 169
sun 12, 39, 49, 67, 76, 79, 91, 97, 109, 118, 162, 166
surface area-to-volume ratio 56
surface tension 2, 17, 36, 39, 49, 53, 59, 76, 87, 94, 95, 120, 121, 125, 128, 154, 162, 165, 168, 170
survival of the fittest 88
swim bladder 2, 10, 11, 21, 27, 40, 48–50, 54, 67, 75, 86, 107, 121, 137, 143, 144, 156, 164, 165
symbiogenesis 2, 3, 5, 6, 9, 22, 23, 27–30, 33, 50, 53, 54, 56, 58–60, 62, 63, 67, 68, 77, 78, 83, 87, 88, 103, 105, 112, 115, 124, 125, 128, 129, 133, 143, 144, 153, 155, 157, 163, 168, 169

T

Target of Rapamycin 87, 88, 154, 161
Teilhard de Chardin 127
teleology 86
Terminal Addition 23, 26, 40, 62, 63, 91, 113, 119–123, 131, 156
Thomas Berry 127
time 4, 6, 11, 21–23, 31, 33–35, 42–45, 63, 71, 75, 82, 83, 87, 91, 99–101, 104, 108, 114, 116–118, 121, 123, 124, 132, 133, 142–144, 147, 148, 150–152, 155, 166, 170
topology 2, 65, 68, 70, 71, 80, 82, 94, 95, 97, 99, 101, 162, 166
trachea 31, 48, 49, 75
Trefoil Knot 161

U

unconscious 95, 141, 142
unicell 5, 8, 13, 16, 18, 29, 3, 44, 50, 53, 54, 70, 78, 82, 83, 86, 87, 90, 95, 99, 104, 105, 109, 125, 155, 166

V

Varela 47
vertebrate 22–24, 27–29, 33, 36, 39–41, 48, 49, 53, 67, 81, 87, 96, 98, 107, 113, 115, 116, 119, 122, 123, 125, 131, 137, 138, 139, 144, 146, 148, 154, 164

W

wave collapse 17, 51, 78, 83, 100, 142, 170
Wolpert 28, 47, 151

Y

Yeast 26, 61, 130, 154, 155

Z

zygote 17, 29, 44, 47, 79, 83, 96, 100, 119, 122, 123, 143, 161, 163

For Product Safety Concerns and Information please contact our EU
representative GPSR@taylorandfrancis.com
Taylor & Francis Verlag GmbH, Kaufingerstraße 24, 80331 München, Germany

* 9 7 8 1 0 3 2 1 9 7 0 2 9 *